PRINT COMPANION

for

Earth Matters

Environmental Geology CD-ROM

Jeremy Dunning
Larry Onesti

Indiana University, Bloomington

W. H. Freeman and Company
New York

DIRECTOR OF ELECTRONIC MEDIA: *Patrick Shriner*
PROJECT EDITOR: *Erica Seifert; Marci Nugent, Editorial Services of New England*
ADMINISTRIATIVE ASSISTANT: *Ceserina Pugliese*
COVER DESIGN: *Maureen Friel, Scott Routen, Artifex, Inc.*

ISBN: 07167-3101-0

Printed in the United States of America

First Printing, 1998

Contents

Preface

*T*his booklet is the **Print Companion** for the *Earth Matters: Environmental Geology CD-ROM*. The companion, the CD packaged herein, and a web site that has been established constitute the three components of the *Earth Matters* learning package.

At the core of *Earth Matters* are 37 lessons on topics relating to environmental geology. These lessons, called *units*, are designed as supplemental learning tools for a college-level geology course. It can be used in any course where a focus on environmental topics is desired. The 37 unit lessons are divided into eight *modules* of broader topics.

As a supplement, *Earth Matters* presupposes that the user will have access to more in depth information about most of the topics covered in the unit lessons, either through a textbook or an instructor's lectures. *Earth Matters* is an interactive electronic learning tool, and therefore is designed to give users a different type of learning experience than they will get from books or lectures. It emphasizes hands-on "learning science by doing science." Through interactivity, the CD lets the user make choices and set variables and then graphically see the consequences of these choices.

In the Print Companion

*T*his **Print Companion** has been developed as an aid to help students and instructors get the most out of the CD. Specifically, it has been designed with these potential uses in mind:

- An introduction to the interactive materials on the CD that users can read before and/or refer to during their use of the CD
- A guide to studying with *Earth Matters*, with suggested learning objectives offered for each unit lesson on the CD
- A workbook for using *Earth Matters* as a lab, with a short lesson sheet offered for each of the 37 units on the CD
- Similar to above, a means for assigning the unit lessons on the CD as homework
- A way to correlate the lessons on *Earth Matters* to the coverage of topics presented in two Freeman geology textbooks—*Environmental Geology: An Earth System Science Approach* by Dorothy Merritts, Andrew De Wet, and Kirsten Menking; and *Understanding Earth*, Second Edition by Frank Press and Raymond Siever

After the unit lesson sheets, starting on page 95, there are three additional features in this companion.

- A glossary of important geological terms used on the CD and in the companion
- Answers to all the questions posed in the unit lessons
- Lists that correlate the units of the CD to the Merritts, DeWet, and Menking textbook and the Press and Siever textbook from the perspectives of those books

On the CD-ROM

*T*here are numerous features on the CD that provide users with an engaging study of its topics. Here are descriptions of some of the more important ones:

- "Reality-based" animations are visual-pleasing and scientifically-accurate graphic representations of key geologic processes. Each is based on real scientific data. There is list of animation topics in the "Special Features" section after this Preface. Please note that animations will run can be viewed full screen with only a slight degradation of the image by clicking the cursor on the animation while it is running. To return to the normal size, simply click on the image again.
- QuickTime™ Virtual Reality (QTVR™) field trips of significant geological sites. These too are listed and described in the "Special Features" section.
- Capstone units appear at the ends of six of the eight modules. These are exercise-based and cover the topics presented in other units of the modules in which they appear.

The units on the CD have similar design and pedagogical conventions. However, no two are exactly alike in either look or functionality. Each lesson has been developed to reflect the uniqueness of the topic being covered. It is also hoped that the variety of learning experience will keep the user more engaged as they study.

On the Web Site

*E*arth Matters has a web site that adds even more functionality to the CD. The web site offers continually-growing environmental coverage not found on the CD or in the companion—including ten full lab experiments which are more comprehensive than the 37 unit lessons found in the companion. The web site can be found at this address:

www.whfreeman.com/geology

There is list of web site's full lab experiment topics in the "Special Features" section after this Preface.

To the Student

*T*he *Earth Matters* CD-ROM can help you succeed in your study of geology. The content of *Earth Matters* is designed to help you in several ways: to clarify concepts, to help you organize study sessions, and to save you time in the lab. What makes this CD-ROM invaluable to you, the geology student, is that with it, you will be able to see geological changes occur. The same geological transformations that sometimes can be tedious to read about in formal prose can be quite spectacular to observe.

The best way to approach a unit on the CD-ROM is to first read its brief introduction to it in this **Print Companion**. The introductions are set up to give you a quick idea of the themes of each unit (the learning objectives), to provide background on the ideas around which the activities of the unit on the CD-ROM revolve, and to pose a few questions to help you review the unit. It is our hope that you find *Earth Matters* informative as well as fun to learn from. Good luck!

List of Special Features

Animations

Earth Matters contains numerous original animations depicting geological processes. There are 26 major animations, with some having more than one part. Additionally, there are numerous "morphs" (short for *metamorphosis*, meaning transformation) showing rock materials changing from one form to another in the Materials and Processes module. Here is a list of the animations and morphs and the units in which they appear.

Module One: Materials and Processes
Unit 2: Igneous Rocks
- Igneous Rock Cooling—Rapid, Slow, and Mixed
- Several Rock Morphs

Unit 3: Metamorphic Rocks
- Several Rock Morphs

Unit 5: Rock Cycle
- Weathering
- Deposition and Burial
- Heat and Pressure
- Melting
- Cooling
- Uplift

Unit 9: Capstone for Materials and Processes
- Several Rock Morphs

Module Two: Earthquakes and Volcanoes
Unit 1: Plate Tectonics
- Continent-Continent Convergent Boundary (Himalayan Orogeny)
- Continental Divergent Boundary (Red Sea Rift)
- Ocean-Continent Convergent Boundary (Subduction along Andes Mountains)
- Transform Boundary (San Andreas Fault System)

Unit 2: Stress States
- Normal Fault Earthquake
- Reverse Fault Earthquake
- Strike-Slip Fault Earthquake

Unit 3: Magma Types
- Shield Volcano/Caldera formation
- Composite Volcano
- Cinder Cone Volcano

Module Three: Surface Processes
Unit 1: Rivers
- Evolution of a River

Unit 2: Wind
- Wind and Sand Dune Types

Unit 3: Coastal Processes
- Wave Refraction
- Longshore Drift
- Rip Current

Unit 4: Glacier
- Mass Balance of a Glacier — Ablation, Accumulation, Equilibrium

Module Five: Natural Hazards
Unit 3: Flooding
- Drainage Network Flooding — Convectional Storms
- Drainage Network Flooding — Frontal Storms
- Drainage Network Flooding — Frontal Storms on Snowpack

Videos

A few video clips of surface processes are offered on *Earth Matters*. These are listed below along with the unit in which the appear.

Module Five: Natural Hazards
Unit 1: Earthquakes
- Liquefaction
- Earthquake Scenes
- Earthquake Damage Scenes

QuickTime™ Virtual Reality (QTVR™) Scenes

*A*pple Computer's QTVR™ software has been employed in a number of scenes in three settings to allow users to take "virtual field trips." Here is a list of the settings and scenes.

Module Eight: Field Trips
Unit 1: Landfills
Setting: Monroe County Sanitary Landfill (Indiana)
- Area 1: Burial of waste material
- Area 2: Leachate collection
- Area 3: Creation of landfill, first part
- Area 4: Creation of landfill, second part

Unit 2: Faults
Setting: In and around Hayward, Califorinia[1]
- Area 1: Hayward fault in rural area
- Area 2: San Andreas fault at reservoir
- Area 3: Hayward fault in urban setting

Unit 3: Tectonics
Setting:
- Area 1: Plate convergence, first part
- Area 2: Plate convergence, second part

Web Site Laboratory Experiments

A special feature of the *Earth Matters* project is a collection of ten full environmental geology lab experiments found on the web site. These are more comprehensive than the review lessons found in this **Print Companion**. They can be easily downloaded or printed directly from the web site. For each lab experiment, users should go through the material on *Earth Matters* that relates to the topic, study the additional material in the lab, and then complete the exercises that conclude the labs. Here is a list of the lab experiment topics.

1. Rocks: Origins and Evolution
2. Volcanic Eruption and Public Policy Planning
3. Earthquake Analysis
4. Rivers, Erosion, and Weathering
5. Mass Wasting
6. Plate Tectonics
7. Groundwater
8. Waste Disposal
9. Pollutants: Survey and Analysis
10. Energy Resources

---

[1] The authors and publisher wish to expend a special thanks to Dr. Sue Hirschfeld of California State University, Hayward for her assistance in filming the Hayward scenes.

Module One
Materials and Processes

Module One deals with the building blocks of geology—the materials and processes that make up the system in which we live. Your ability to learn about the more specifically environmental topics come after Module One depend on you have a firm grasp of the underlying ideas presented in this module.

The first unit examines the earth's minerals. There are over 3,000 minerals, but only about 100 are found in abundance. The minerals can be placed into three chemical groups: silicates, oxides, carbonates, and sulfides and sulfates. These three types of minerals are the constituents, either alone or in combinations, of all the earth's rocks.

All rocks are formed out of magma and thus in one way or another are *igneous rocks*. Some escape metamorphosis and sedimentary processes and remain pure igneous rocks. That group is the subject of the Unit #2. *Metamorphic rocks*, the subject of Unit #3, are igneous rocks that are metamorphosed in the earth's interior from one solid form to another solid form, without passing through a molten stage. Unit #4 covers *physical properties* of rocks—those characteristics of rocks that dictate their response in various states of stress (whether they break into fragments or remain intact and fold).

Unit #5 explores the *rock cycle* and its rock-transforming processes. This leads to the last of the major groups of rocks: the *sedimentary rocks*, which evolve as products of loose igneous and metamorphic rock fragments. The next two units examine the characteristics and durability of *soils*, which are made up of weathered rock materials, and the effects of chemical and physical *weathering* on the earth's surface materials. The final unit is a "capstone" exercises that calls upon knowledge you have acquired.

Here is one of the important concepts you can take from this module: Nothing in, on, or above the earth is static. Rather, the earth is a complex system of systems in which little or nothing is wasted and all systems drive other systems. Therefore, actions or events that seemingly have no connection to each other may indeed impact on each other. These ideas are at the heart of the "earth system science" approach—a way of looking at geology that works very well with an environmental focus.

Units

1. Mineralogy

The types of minerals, their properties and structure, and the rocks in which they occur

2. Igneous Rocks

The types of igneous rocks that result from eruptions of different magma types cooled at various rates and the types formed in different plate tectonic settings

3. Metamorphic Rocks

The effects of different types of metamorphism on sandstone, shale, diorite, andesite, granite, rhyolite, gabbro, limestone, arkose, and slate

4. Physical Properties

The effects of compressional, tensional, and shear stress on strong and weak rock materials

5. Rock Cycle

The effects on rocks of the continuous cycle of weathering, deposition and burial, heat and pressure, melting, cooling, and uplift

6. Sedimentary Rocks

The origin and properties of chemical and clastic rocks

7. Soils

The types of soils that develop in different time frames, climates, and vegetation settings from different parent materials on various slopes of land

8. Weathering

The types and effects of chemical and physical weathering agents in midlatitude, warm arid, warm tropical, and polar climates

9. Capstone for Materials and Processes

An exercise in rock forming

1. Mineralogy

Minerals are the building blocks of the earth.

Learning Objectives

- To be able to list the four basic types of minerals

- To be able to describe the chemical makeup of the four basic minerals

- To be able to list the major characteristics geologists use to classify minerals

- To be able to identify the class of minerals found in igneous, sedimentary, and metamorphic rocks

Introduction

*L*inguists work with alphabets to reduce language to its simplest form to enable them to analyze its structure and components. In a similar way, geologists work with four mineral types that they have discovered to be the most basic constituents of all rocks: carbonates, silicates, oxides, and sulfates and sulfides.

Geologists define a *mineral* as a naturally occurring, solid crystalline substance, generally inorganic, with a specific chemical composition. *Crystalline* means that the atoms that compose the mineral are arranged in an orderly, repeating, three-dimensional array. The quality of being *generally inorganic* means that the materials such as coal that are made up of decaying vegetation and animal remains are not generally considered minerals. The *chemical composition* of minerals is fixed or varies within narrow limits. Quartz, for example, has a fixed ratio of two atoms of oxygen to one of silicon.

Geologists use their knowledge of minerals to gain understanding of the earth's history. The types of rocks found around some volcanoes, for example, informs researchers of the types of processes that must have occurred there. Knowing the ranges of temperatures and pressures required to produce a certain mineral or to change its form from one mineral to another allows researchers to deduce the nature of the geologic events that had to have occurred at a particular site. It has been this knowledge of minerals as basic building blocks of the earth that has helped scientists piece together a geological history of the earth and to develop the theories of continental drift.

This unit on *Earth Matters* examines the four key aspects of mineralogy: The four major types of minerals, the major properties by which minerals can be identified and classified, the chemical structure of minerals, and the mineral composition of the major rock types.

Review

1.   Give an example of a mineral in the oxide group.

_____

_____

2.   What is *luster*?

_____

3.   "One atom of carbon and three atoms of oxygen ions" describes which type of mineral?

_____

4.   In the earth, what class(es) of minerals tend to make up igneous rocks?

_____

2. Igneous Rocks

*Igneous rocks are formed from magma extruded
onto the surface or intruded into the crust of the earth.*

Learning Objectives

- To understand the various ways geologists classify magmas and rocks formed from them

- To be able to describe the texture of a rock that would result from different cooling rates of magma

- To be able to describe the effect of plate tectonic settings on the formation of igneous rocks

Introduction

The four basic types of minerals—carbonates, silicates, oxides, and sulfides and sulfates—are combined and recombined by earth's processes into three major families of rocks: igneous, sedimentary, and metamorphic. Igneous rocks, the subject of this unit, are formed out of the earth's magma.

Igneous rocks form from the crystallization of magma. When magma cools slowly inside the earth to a temperature just below the melting point so that the rock remains solid, large crystals form, and a coarse-grained igneous rock results. But when magma erupts onto the surface of the earth, it cools very rapidly, and the crystals that forms are small, resulting in a fine-grained igneous rock.

Geologists distinguish two major types of igneous rocks based on the size of their crystals: intrusive and extrusive. *Intrusive igneous rocks* are formed by slowly crystallizing magmas that have intruded rock masses deep in the interior of the earth. They can be recognized by their interlocking large crystals, which grew slowly as the magma gradually cooled. Magmas cool slowly when they invade rock masses because the rock mass acts as an insulator that maintains the heat in the invading magma. *Extrusive igneous rocks* form from magma that cools rapidly upon reaching the earth's surface. Extrusive igneous rocks are easily recognized by their glassy or fine-grained texture. Eruptions of magma through volcanism produce almost instantaneously crystallized ash particles that rise quickly to the surface and flow as liquid for some distance before they solidify.

The three types of magma are distinguished by viscosity and silica content: Basaltic magma is composed of about 50 percent silica and has a low viscosity; andesitic magma is composed of about 60 percent silica and has a medium viscosity; and rhyolitic magma is composed of about 70 percent silica and has a high viscosity. The cooling history of a particular magma determines the type of rock that will be formed from it.

Basaltic magma forms mafic minerals. Andesitic magma forms minerals in the intermediate group. Rhyolitic magma forms felsic minerals.

There are several animations and "morphs" in this unit designed to help you see the igneous rocks that result from combining different cooling histories with the three types of magma in different tectonic plate settings.

Review

1.   When magma erupts onto or near the earth's surface and cools rapidly, what type of rock texture will the cooled rocks have? What type of texture do they have if the magma intrudes into deep rock inside the earth?

_____

2.   What island chain is offered as an example of a site of rapidly-cooling mafic magma?

_____

3.   In rift zones at divergent plate boundaries, what often happens to the magma that doesn't reach the surface and erupt in a basaltic flow? (see "mafic," "slow")

_____

3. Metamorphic Rocks

Metamorphic rocks are the result of forces in the earth that restructure existing rocks.

Learning Objectives

- To be able to predict the new form an existing rock will take after having been subjected to a change agent—regional, dynamic, or contact metamorphism—to varying degrees—high, medium, or low

- To be able to associate metamorphic processes with different types of plate tectonics

Introduction

Metamorphic rocks are produced deep in the earth, where high temperatures and pressures cause their mineralogy, texture, or chemical composition to change while they remain in a solid state. These rocks, consisting mostly of silicates, can be igneous, sedimentary, or other metamorphic rocks.

Metamorphism can occur over a large area, in which case it is referred to as *regional metamorphism*. If it occurs over a relatively small area close to the surface of a fault, it is referred to as *dynamic metamorphism*. If it occurs in or near the area of an igneous intrusion, it is referred to as *contact metamorphism*.

Regional metamorphism occurs mainly in convergent plate boundary areas where an oceanic plate is subducted beneath another oceanic plate or a continental plate. Dynamic metamorphism occurs only in fault zones, which develop in transform and convergent plate boundary areas. Contact metamorphism occurs in only those plate boundary areas where magma intrudes into surrounding rock—that is, in the marginal areas between divergent or convergent plates.

This unit of *Earth Matters* on the CD will enable you to see animated simulations of numerous examples of metamorphism. You choose the starting rocks and then the type of metamorphic process. After making these selection you click on the "See It" button for a description and "morph" of the process.

Review

1.   What three factors cause regional metamorphism?

_____

_____

_____

2.   What causes rocks to change in a dynamic metamorphic situation?

_____

_____

3.   How do intrusive igneous rocks cause contact metamorphism to occur?

_____

_____

4. Physical Properties

The physical properties of a rock control the behavior of the rock.

Learning Objectives

- To be able to explain the difference between brittle and ductile rock behaviors

- To be able to explain how the properties of a rock make it strong or weak

- To be able to predict the behavior of ductile and brittle rocks in different stress states

- To be able to associate types of stress states with plate tectonics

Introduction

*H*ow rocks respond to stress, that is, whether they burst apart or fold up, depends on their strength and on the type and strength of the stress they are subjected to. *Compressional stress* pushes rocks together, which is the direction of the stress that occurs at convergent plate boundaries. *Tensional stress* results from the earth's plates pulling away from each other, as happens in divergent plate boundary areas. *Shear stress* results from the earth's plates moving side by side, as happens in transform plate boundary areas.

If a strong rock is brittle and subjected to compressional stress, it will behave in a linear, elastic fashion until the elastic limit it reached, at which point it will fail by fracture. When the fracture occurs, large amounts of stored elastic strain energy will be released in the form of microseismic or seismic waves. If, in contrast, a strong rock is ductile and subjected to compressional force, it will display linear elastic behavior until the elastic limit is reached, at which point the stress-strain relationship will flatten out while flow occurs within the material.

*T*he graphics in this unit on *Earth Matters* will help you evaluate the effects of various factors on the metamorphosis of rocks. You are called upon to set up stress situations by choosing between levels of rock strength, brittle or ductile behavior, and stress state. After you set these conditions you can see the results on a graph.

Review

1.   Which stress state tends to promote sliding along a plane?

_____

2.   When brittle rock fractures, what happens to the stored elastic strain energy?

_____

3.   If a strong material behaving in a brittle fashion were subjected to shear stresses, what type of fracture would result?

_____

4.   If a strong material behaving in a ductile fashion were subjected to shear stress, what would the effect be on the surrounding shear zone?

_____

5. Rock Cycle

Rocks evolve with time and change form through the rock cycle.

Learning Objectives

- To be able to describe the six processes in the rock cycle—weathering, deposition and burial, heat and pressure, melting, cooling, and uplift

- To be able to explain the physical changes that occur in each of the six processes of the rock cycle

Introduction

The *rock cycle* is a set of geologic processes by which each of the three great groups of rocks—igneous, sedimentary, and metamorphic—is formed from the other two. Subjected to weathering and erosion, rocks form sediments, which are deposited, buried, and lithified. After deep burial, the rocks undergo metamorphism, melting, or both. Through mountain building or volcanism, the rocks are uplifted, only to be recycled again.

The rock cycle is only a sequence of processes; the amount of time rocks spend in any part of the cycle varies considerably, and there is no beginning or ending point, just a continuous cycle of change. For the purposes of description, however, we can explore the cycle as if it began in the earth's interior.

Rock begins as molten material, magma, deep inside the earth where temperatures are greater than the melting point of most rock, 700° centigrade. Geologists call this phase the *plutonic episode*, named after the Roman god of the underworld, Pluto. At this place, all preexisting rocks are melted, and all their component minerals are destroyed and their chemical elements homogenized in the resulting hot liquids.

When the magma reaches the surface due to faulting or volcanism at divergent or convergent plate boundaries, it cools and begins to crystallize. During the ensuing mountain building, rocks are uplifted, where they are exposed to weathering and erosion forces and consequently form sediments. From the elevation of the mountain, water and wind carry the sediments down to the oceans and continents, where they are buried and lithified, to become sedimentary rocks. Sediments that reach the ocean floors form increasingly heavy layers that eventually plunge deep into the earth as oceanic plates slowly move away from the midocean ridges.

As the lithified sedimentary rock is buried more deeply in the crust, it becomes hotter. As the depth of burial exceeds 10 km and the temperatures climb, the minerals in the still-solid rock start to change to new minerals that are more stable at the higher temperatures and pressures of the deeper part of the earth's crust. This is the process of metamorphism that transforms sedimentary rocks into metamorphic rocks. With further heating, the rocks may melt and form a new magma from which igneous rocks will crystallize, starting the whole cycle again.

Each of the six parts of the rock cycle is represented with an animation designed to help you better understand that part. Be sure to listen the narration as well as reading the information box for each animation.

Review

1.   "Divergent boundary" is offered as an example of a plate tectonic setting for which part of the rock cycle on the CD?

   _____

2.   What is lithification? (see "Deposition/Burial")

   _____

   _____

3.   How are rocks changed in uplifted areas exposed to weathering?

   _____

6. Sedimentary Rocks

Sedimentary rocks are formed by the effects of wind, water, ice, and gravity.

Learning Objectives

- To be able to describe the formation process of chemical sedimentary rocks

- To be able to describe the formation process of clastic sedimentary rocks

Introduction

*S*edimentary rocks are assembled out of the earth's sediments, which comprise a rather eclectic set of mismatched particles of silt, sand, and shells of organisms, as well as fragments of igneous and metamorphic rocks that have been broken down by weathering. Scientists group these particles and remains of weathering and erosion into *clastic sediments*, that is, sediments that are physically laid down by running water, wind, and ice, which then form layers of sand, silt, and gravel, and *chemical and biochemical sediments*, that is, sediments that are formed of precipitates of chemical weathering processes.

All sedimentary rocks are formed by the layering of sediments. There are two means by which loose sediments are transformed into solid rock: *compaction,* that is, the pressing together of sediments by the weight of overlying sediment, which creates a mass that is denser that the original sediments, and *cementation*, that is, the binding together of sediments by minerals that precipitate around the loose sediments, which they eventually glue together.

*T*he CD describes the types of sedimentary rocks—chemical (and biochemical) and clastic. It further describes classifications made within these two types of sedimentary rocks.

Review

1.   How is rock salt formed?

_____

2.   How do organisms form sedimentary materials?

_____

_____

_____

3.   What characteristic distinguishes glacial sedimentary rock deposits from other
        types of transported materials?

_____

7. Soils

The breakdown of rock material and organic matter results in the creation soils.

Learning Objectives

- To be able to describe the changes in soil makeup that occur as it ages

- To be able to describe the changes in soil that occur with changes in climate and vegetation

- To be able to explain the results of chemical and physical weathering of parent rock material in the formation of soil

Introduction

During that phase of the rock cycle in which igneous, metamorphic, and sedimentary rocks come to rest upon the surface of the earth, they are subjected to the earth's surface phenomena through which they may be crushed, ground, transported, and sorted out. If the rock material is of a sufficiently small particle size, if it lands in an area in which there is enough water to support living organisms, and if it contains the minerals that the organisms in that place need to survive, it may eventually become soil. Considering the amount of time it takes for all of these processes to play out fully, one wonders that it is ever formed at all.

The existence and makeup of soil depends upon a most complex array of factors because it is part mineral and part organic. Furthermore, seemingly minute variations in any one of those factors—the climate, the minerals in the parent rock material, the slope of the land, or the diversity of the vegetation—can have quite dramatic effects on the ability of the soil to support vegetation, which, of course, supports animal life.

This unit on *Earth Matters* contains a wealth of information broken down into four factors that have great influence on the formation of soils—time, climate/vegetation, parent materials, and slope.

Review

1. Where is the O horizon found in a soil profile?

    _____

2. What would the organic matter and mineral content be in the soil of an arid region?

    _____

3. What property of clay makes its presence an impediment in the soil-forming process?

    _____

8. Weathering

Weathering results in the physical and chemical breakdown of rocks.

Learning Objectives

- To be able to estimate the effect of various climates on the physical weathering of granite, basalt, limestone, shale, sandstone, gneiss, and slate

- To be able to estimate the effect of various climates on the chemical weathering of granite, basalt, limestone, shale, sandstone, gneiss, and slate

Introduction

The temperature, humidity, and winds of a particular region are the most influential factors in the determination of the nature and pace of physical and chemical weathering that occurs there. Soil forms poorly anywhere that is exposed to strong winds or where there is little water. Rocks break up at a faster rate where temperatures run to extremes between freezing and thawing than where temperatures remain in a moderate range. Chemical changes in rocks occur faster in water than they do on land because water, known as the *universal solvent*, can break down the chemical bonds of minerals by dissolution, oxidation, and hydration.

The CD offers explanations of chemical weathering, physical (or mechanical) weathering, and the importance of climate in weathering. You can then see graphic representations of these factors at work by choosing rocks, choosing a climate type, setting the level of weathering, and then clicking on the "See It" button. Side-by-side representations of different rock types are designed to get you thinking about why different rocks react differently to the same conditions.

Review

1. What effect do frequent freezing and thawing periods have on rocks?

   _____

2. What types of weathering result from wind and gravity in areas where surface water is present?

   _____

3. What two factors inhibit chemical weathering in polar climates?

   _____

4. What factor inhibits chemical weathering in arid climates?

   _____

5. What two climatic factors affect the rate of weathering of rocks?

   _____

9. Capstone for Materials and Processes

An exercise in rock forming

Introduction

*A*s you work through these exercises on rock forming, you will have a chance to build some rocks, shaping them and reshaping them, and then seeing what happens to them when they are exposed to the earth's processes. Completing this unit on *Earth Matters* should cement into place any remaining loose information you may still have about the most basic structural unit of the earth's surface: rocks.

Correlation of Module to Freeman Textbooks

EG: *Environmental Geology* by Dorothy Merritts, Andrew De Wet, Kirsten Menking
UE: *Understanding Earth*, Second Edition, by Frank Press and Raymond Siever

Overall Module

EG: Most of the coverage in this module relates to coverage found in Chapters 4 and 6 of the Merritts textbook.

UE: Like Part 1, "Understanding the Earth System" (Chapters 1-10) of *Understanding Earth*, this module is designed to acquaint or re-acquaint students with the basic physical factors of geology. These factors are crucial to an understanding environmental geology just as they are for physical geology.

1. Mineralogy

EG: "Lithosphere Materials: Elements, Minerals, and Rocks" in Ch. 4, pp. 93-107

UE: Chapter 2: Minerals, pp. 26-57

2. Igneous Rocks

EG: "The Rock Cycle" in Ch. 2, pp. 53-55; "Major Rock Groups and the Rock Cycle" in Ch. 4, pp. 100-101; "Distribution of Rock Types" in Ch. 4, pp. 111-118

UE: Chapter 4: Igneous Rocks, pp. 74-103

3. Metamorphic Rocks

EG: "The Rock Cycle" in Ch. 2, pp. 53-55; "Major Rock Groups and the Rock Cycle" in Ch. 4, pp. 100-101; "Distribution of Rock Types" in Ch. 4, pp. 111-118

UE: Chapter 8: Metamorphic Rocks, pp. 194-215

4. Physical Properties

EG: This unit relates most closely to "Earthquakes" in Ch. 5, pp. 146-152.

UE: "How Rocks Become Deformed" in Ch. 10, pp. 247-248

5. Rock Cycle

EG: "The Rock Cycle" in Ch. 2, pp. 53-55; "Major Rock Groups and the Rock Cycle" in Ch. 4, pp. 100-101; "Distribution of Rock Types" in Ch. 4, pp. 111-118

UE: "The Rock Cycle," "Plate Tectonics and the Rock Cycle," pp. 68-71

6. Sedimentary Rocks

EG: "The Rock Cycle" in Ch. 2, pp. 53-55; "Major Rock Groups and the Rock Cycle" in Ch. 4, pp. 100-101; "Distribution of Rock Types" in Ch. 4, pp. 111-118

UE: Chapter 7: Sediments and Sedimentary Rocks, pp. 162-193

7. Soils

EG: "Where Soils Form" in Ch. 2, pp. 42-44; Chapter 6: Soils and Weathering Systems, pp. 158-190

UE: "Soil: The Residue of Weathering" in Ch. 6, pp. 152-158

8. Weathering

EG: "Where Soils Form" in Ch. 2, pp. 42-44; Chapter 6: Soils and Weathering Systems, pp. 158-190

UE: Chapter 6: Weathering and Erosion, pp. 134-161

9. Capstone to Materials and Processes

Both: This unit is an exercise related to the coverage of the other units in the module.

✓   *See the web site for URL links to sites that relate to the topics in this module.*

www.whfreeman.com/geology

Module Two
Earthquakes and Volcanoes

*T*he units in this module on *Earth Matters* explore the causes and effects of earthquakes and volcanoes. Units in a later module—Natural Hazards—explore these events more from the stand point of human and public policy issues. Earthquakes and volcanoes are the most studied natural disasters. Both phenomena are strongly associated with plate tectonics and faults.

The first unit, Plate Tectonics, encapsulates the most current theory and viable explanation of the formation of the earth's surface structures and geological phenomena. Since there is a close connection between plate tectonics and earthquakes and volcanoes, the information in this module is especially important.

The second unit, Stress States, relates to earthquakes. The unit is structured in such a way that you can simulate faults in either brittle or ductile regimes by combining one of the three states of stress with one of the three levels of stress drops. The third unit, Magma Types, illustrates how different types of magma eruptions build up different types of volcanic mountains. The Explosivity unit shows the sometimes catastrophic effects of volcanism.

The capstone reviews earthquake and volcanic activity as their characteristics change with their location on or between the earth's plates. It tests your understanding of the lessons in the other units of this module.

Units

1. Plate Tectonics

Convergent, divergent, subduction, and transform boundaries

2. Stress States

Effects of high, medium, and low levels of stress drops in compressive, tensional, and shear stress state areas in brittle and ductile regimes

3. Magma Types

Magma types--basaltic, andesitic, and rhyolitic—described by viscosity and silica content and consequent volcanic mountain-building properties

4. Explosivity

Effects of convergent and divergent plate boundary area volcanism on air quality, public safety, climate, and geothermal energy production

5. Capstone for Earthquakes and Volcanoes

An exercise in relating the type of earthquake or volcanic eruption to the plate tectonic setting

1. Plate Tectonics

Plate tectonics is the driving force behind many of the earth's processes.

Learning Objectives

- To be able to explain how the study of plate tectonics helps geologists understand some of the events that led to the formation of many of the earth's features

- To be able to explain how volcanoes and earthquakes form during the shifting of the earth's plates

Introduction

The position and symmetry of the earth's land masses has baffled many generations of scientists. Where did that island and that volcano come from? How did this long mountain chain form? Why does it look like the continents could be fit together like the pieces of a jigsaw puzzle? Is it possible that the continents are broken fragments, complete with jagged edges, of what was once one large continent?

Many theories were tried out to explain the evolution of the earth's surface into its present form. In 1912 a German scientist, Alfred Wegener, proposed that 200 million years ago the continents were actually one large continent called *Pangaea* which began to break up and the pieces of it drifted away to form the continents. Other scientists rebuffed the *continental drift theory*, saying that there was neither evidence that continents were free to move nor that there was any force to push or pull them along.

Evidence to support the existence of the two prerequisites remained hidden until the 1960s when Harry H. Hess of Princeton University proposed the theory of *seafloor spreading*, which explains where the forces (*tectonics*) may have come from to break up *Pangaea* and relocate its pieces (*plates*). According to the theory of seafloor spreading, magma from the earth's mantle rises to the earth's surface and is released through a crack in the seafloor, called an *oceanic ridge* or a *midocean ridge*, that has resulted from the divergence of two of the earth's plates. Initiated by convection forces, the continuous eruption and subsequent mounding of volcanic material from this crack provides some of the force necessary to keep the earth's plates in constant motion.

As new crust material is formed and added to the earth's plates, it pressures and pushes the existing plates so that they sometimes break up into smaller plates. Plates may collide, and if one of the plates is heavier than the other, the heavier one will slide under the lighter one. The heavier plate may actually disappear into a trench through which the plate is reabsorbed into the inner layers of the earth. The lighter one may lift up and become a mountain system. Sometimes plates slide against each other along a fault line.

The development of the theory of plate tectonics was a tremendous event in the science of geology. It explains a great many geological events that affect the earth's surface formations. With this theory, scientists have a new mental laboratory in which to advance their understanding of the interrelated phenomena of physical geology.

As you work through the unit on *Earth Matters*, you will see animations that depict some of the effects of plate tectonic activities: convergence, divergence, subduction, and transform.

Review

1.   The Appalachian Mountains are an example of a plate collision at what type of plate boundary?

_____

2.   What is the area called when plates pull apart from each other at divergent boundaries?

_____

3.   Name a geologic process that might occur at a subduction zone.

_____

4.   What type of tectonic plate activity has caused the San Andreas Fault system to occur in southwestern California?

_____

2. Stress States

The deformation of the earth is related to the stress states present at the plate boundary areas.

Learning Objectives

- To be able to make a reasonable estimate of the effects of stress and strain on brittle and ductile rock materials

- To be able to analyze different combinations of stress, strain, and material strength to describe faults, earthquakes, and volcanoes

Introduction

Stress is force or pressure; a *stress state* is the existence of force between two objects. Geologists examine stress conditions at sites of the earth's geologic formations to explain events that have happened or to predict events that will happen.

Compressive stress states exist between two rock formations or two plates that are converging. *Tensional stress states* exist between two rock formations or two plates that are diverging. *Shear stress states* exist between two rock formations or two plates that are sliding laterally past each other. When a state of stress exists between two rock formations or two plates, the effect of the force will be determined partly by the nature of the material making up the boundary area of the plates. *Brittle material* tends to break apart whereas *ductile material* tends to fold or bend.

The degree of stress and the amount of strain, together with the makeup of the material, will determine the consequences of fault, volcanic, and earthquake activity.

*O*n the CD you can see the effects of various features of stress, by with different theoretical combinations of stress, strain, and rock material. You select among stress states and stress drops (strains) and then chose between brittle and ductile rock materials. Clicking on the "See It" button give you a representation of the results of the combination you have chosen—either with an animation, a photograph, or the message that the combination you have chosen is not applicable.

Review

1. In what type of tectonic activity do tensional stresses occur?

_____

2. In a brittle regime, high-stress-drop situation, what three forms of elastic strain energy are released?

_____

3. In a brittle regime, what types of faults result from compressive stresses?

_____

4. In a normal fault, in what direction does the hanging-wall block move in relation to the footwall block?

_____

5. Why does the ductility of rock increase with depth in the earth's crust?

_____

_____

3. Magma Types

Volcanoes are the result of the eruption of various types of magma (molten rock).

Learning Objectives

- To be able to describe the three basic types of magma—*basaltic, andesitic,* and *rhyolitic*—in terms of their respective viscosity and silica content

- To be able to associate the three basic types of magma with their mountain-building properties

Introduction

Volcanism occurs when molten rock in the earth's mantle rises up and through the earth's crust. The mantle is a pressurized, fluid and gaseous system that is confined by the shell of the earth's crust. Pressure within the mantle is created and maintained by convection (heat) forces, which expand the molten material and gases. Wherever the earth's shell is ruptured in some way by plate tectonics, there is a potential for molten material from the earth's mantle to explode through it. These explosions are referred to as *volcanism*. The mountain forms that are built up by the material expelled by the explosions are referred to as *volcanoes*. The composition and shape of volcanoes are determined largely by the viscosity and silica content of the magma.

According to plate tectonics theory, the motion of the earth's plates can cause ruptures in the earth's crust through which volcanism may occur in four ways:

1. Where plates diverge, such as at the Mid-Ocean Ridge
2. Where plates converge and a heavy continental plate is subducted beneath a lighter oceanic plate, such as at the Andes mountain system
3. Where two of the earth's oceanic plates collide, causing volcanic island arcs to form such as the Marianas in the western Pacific Ocean
4. Where plates are in motion over areas of the mantle that are particularly hot, such as Yellowstone.

On the CD-ROM, you will be able to view the formation of shield, composite, and cinder cone volcanic mountains as they are built up by, respectively, basaltic, andesitic, and rhyolitic magma flows. You are also given information about the degree of viscosity and the silica content of each magma type.

Review

1. What effect does the rate of cooling have on the crystallization of melted rock material?

_____

2. What is a shield volcano?

_____

3. What is the viscosity of each of the three basic types of magma?

_____

4. What type of volcano results from eruptions of basaltic magma?

_____

5. What type of volcano results from eruptions of andesitic magma?

_____

6. What type of volcano results from the eruptions of rhyolitic magma?

_____

4. Explosivity

Convergent and divergent plate boundary volcanism affect
air quality, public safety, climate, and geothermal energy production.

Learning Objectives

- To be able to describe some of the environmental effects of volcanism

- To be able to describe some of the natural hazards posed by volcanism

- To be able to explain how volcanism can be used as an inexhaustible source of energy for human endeavors

Introduction

Volcanism is one of the most violent of all the earth's geological phenomena, and it naturally follows that the consequences are significant and spectacular. Many human lives have been lost in fast-moving lava flows and explosions of poisonous gases. In 1985 the volcano Nevado del Ruiz in Columbia, South America, was responsible for the deaths of 25,000 people. There are 500 to 600 active volcanoes on the earth's surface that can claim human lives in a variety of ways—explosive blasts, ashfalls, release of poisonous gases, lava flows, and mudflows that are caused by the eruptions. Scientists have devoted a great deal of study and research to the cause of predicting volcanic eruptions so that people can be evacuated in time to save their lives.

Other long-lasting effects follow in the aftermath of volcanic eruptions. For example, carbon dioxide is sometimes released in quantities that can overwhelm the atmospheric chemical balance and lead to air pollution. Particulate matter is often exploded into the air where it forms clouds that filter out sunlight so that climate is actually changed. Not all effects are negative, however. Volcanism can create geothermal energy, which is an inexhaustible source of fuel for human activities.

The effects of volcanism that are caused by different types of volcanoes that are active in plate marginal areas are shown vividly on the CD-ROM. You are prompted to choose between plate boundary types and among effects of volcanism. Each selection provides information about the boundary type or effect. Then you click on "See It" to learn more about the particular combination of type and effect you have chosen.

Review

1.   How can volcanism sometimes cause air pollution?

_____

2.   How can volcanism actually benefit human activity?

_____

3.   How can volcanism sometimes cause changes in climate?

_____

_____

_____

4.   How can volcanism be hazardous to human safety?

_____

_____

_____

30   EARTH MATTERS: ENVIRONMENTAL GEOLOGY CD-ROM

5. Capstone for
Earthquakes and Volcanoes

An exercise in relating the type of earthquake or volcanic eruption to the plate tectonic setting

Introduction

Working through this capstone exercise on *Earth Matters* will help to reinforce your knowledge of plate tectonics. It consists of four exercises—each one focusing on a different plate tectonic setting: transform boundary, divergent boundary, subduction zone, and an island arc subduction zone. In each exercise you are shown the sites of geologic processes and asked to choose from among the ten processes it might be.

Correlation of Module to Freeman Textbooks

EG: *Environmental Geology* by Dorothy Merritts, Andrew De Wet, Kirsten Menking
UE: *Understanding Earth*, Second Edition, by Frank Press and Raymond Siever

Overall Module

EG: Most coverage is in Chapter 5.

UE: Earthquakes are mainly covered the chapters in Part 3 "Internal Processes, External Effects." Volcanoes are the subject of Chapter 5 in Part 1.

1. Plate Tectonics

EG: Chapter 4 Introduction, pp. 92-93; "Plate Tectonics and the Rock Cycle" in Ch. 4, pp. 102-110; "Volcanic Eruptions," pp. 138-139; "Crustal Deformation Associated with Earthquakes in the Cascades Region" in Ch. 5, pp. 153-154

UE: "Plate Tectonics: A Unifying Theory for Geological Sciences" in Ch. 1, pp. 14-20; "The Global Pattern of Volcanism" in Ch. 5, pp. 121-125; The Big Picture: Earthquakes and Plate Tectonics, in Ch. 18, pp. 470-472; Chapter 20: Plate Tectonics, pp. 504-535

2. Stress States

EG: "Earthquakes" in Ch. 5, pp. 146-152

UE: "How Rocks Become Deformed," pp. 247-248

3. Magma Types

EG: "Igneous Rocks" in Ch. 2, p. 53; "Igneous Rocks" in Ch. 4, pp. 111-113; "Volcanic Eruptions" in Ch. 5, pp. 138-146

UE: "How Does Magma Form?" in Ch. 10, pp. 83-84; "Where Does Magma Form?" pp. 85-86; Chapter 5: Volcanism, pp. 104-133

4. Explosivity

EG: "Volcanic Eruptions" in Ch. 5, pp. 138-146

UE: Chapter 5: Volcanism, pp. 104-133

5. Capstone for Earthquakes and Volcanoes

Both: This unit is an exercise related to the coverage of the other units in the module.

 See the web site for URL links to sites that relate to the topics in this module.

www.whfreeman.com/geology

Module Three
Surface Processes

*T*he movement of wind and water over the face of the earth gives its features much of their beauty. Canyons are carved out of solid rock by relentless rivers; high mountains of sand are built up by steady winds. Glaciers pick up, move, deposit, push into shape, and then leave behind entire landforms built of packed sediment.

The first unit in this module, Rivers, explains the calculations for *river discharge*, a term used to describe the strength and power of a river. Rivers shape surface features in two ways: by erosion or dissolution of the rock surfaces over which they flow. In addition, they pick up and carry a great deal of sedimentary material, thus significantly depleting the sediment in some areas and building it up in others.

The second unit in this module, Coastal Processes, explains how water movement affects the nature and shape of the shorelines. At the edge of continents, oceans shape the surface features to the same extent that rivers shape the inland areas, through erosion or dissolution. Oceans also move a great deal of sedimentary material, such as sand, thus depleting the sediment in some areas and building it up in another.

The third unit, Wind, examines the action of wind and its power to wear away rock and to move sediment from one place to another. Finally, the fourth unit, Glaciers, examines the action of snow and ice in developed glaciers in their power to wear away the surface and to move rock and sediment from one place to another. The module concludes with a capstone unit.

The four primary surface processes are enormously powerful within the context of earth's systems.

Units

1. Rivers

Some of the factors that affect the discharge and carrying power of rivers

2. Coastal Processes

Variations in sea level, plate tectonic activity, waves and currents, types of shorelines, and types of shoreline remediation

3. Wind

The relationship between climate and wind, desert landforms, and desert formation

4. Glaciers

Glacier formation and movement and glacier-created landforms

5. Capstone for the Surface Processes

An exercise in placing surface processes within the earth system context

1. Rivers

Rivers and running water are the primary architects of geologic landscapes.

Learning Objectives

- To be able to calculate the discharge of a river at a particular point

- To be able to calculate the effect of increasing or decreasing the discharge of a river

- To be able to gauge the effects of changes in sea level on the discharge of rivers

- To be able to explain the effects of plate tectonics on rivers

Introduction

Rivers and streams sculpt the earth's features out of the rock on its surface, artistically rounding it or cutting it to create waterfalls and rapids in one place, colorful river canyons in another, plush rainforests in yet another. The speed and precision with which they work are determined largely by how fast they move and how steeply they fall.

The movement of water in a river channel is described empirically as the rate of discharge. A river's *discharge* may be computed from the following formula: The cross-sectional area, that is, the amount of water (m^2) times the velocity of the water (m/s) passing by a particular point equals the rate of discharge (m^3).

The rate at which water travels in a streambed determines, among other things, how much sediment it can carry along, where that sediment eventually settles, how fast it erodes areas it passes through, whether it pursues a straight path or meanders along, whether it floods or recedes, and whether it stays together or divides into smaller streams.

Plate tectonics affects river discharge where a river's headwaters (the source of the river) or base-level areas (the body of water into which the river flows) are raised or lowered by faulting or mountain building.

This unit of *Earth Matters* contains a great deal of information. It starts with an introductory section which offers an animation that provides a framework to the rest of unit. It shows describes and depicts various aspects of a river at three key points. Aspects of the animation are described in greater detail in the other sections of the unit. A section on river discharge shows the concept by describing how discharge is measured at three points. Tectonic activity is described in a section showing the effects of mountain building and rifting. Sediment load is also given clearer meaning by being shown at three points in the river. Finally, a section on sea level variation shows the effects on the coast of the raising and lowering of sea level.

Review

1. What type of sediment load is composed of coarser-grained sediment that is too heavy to be suspended?

    _____

2. What effect does mountain building have on river discharge?

    _____

    _____

3. What effect does a lowering of the sea level have on river discharge?

    _____

    _____

2. Coastal Processes

Shorelines are an indicator of tectonic activity, sea-level variation, and erosion and deposition.

Learning Objectives

- To be able to estimate the types of shoreline changes that occur as a result of plate tectonic activity

- To be able to estimate the types of shoreline changes that occur as a result of waves and currents

- To be able to describe some of the mechanical means people use to prevent coastal erosion

Introduction

Coastal areas are often more attractive to human population than the interiors of continents. Since they are also in many ways the areas that are the most fragile and subject to change, the popularity of coastal area as living places can have some significant environmental consequences.

Global warming is currently one of the most debated scientific topics. While no one can yet definitely state whether it is a fact or an exaggeration, no one can deny that population centers in low-lying areas—including Miami and most of the rest of Florida, coastal Bangladesh, much of the Netherlands, New Orleans, and coastal Maryland— would be submerged by sea water if the earth's sea level were to rise by a relatively small amount.

Coastlines are subject to many geologic factors in addition to sea level. These factors include the types of waves and currents that shape shorelines as well as the pervasive influence of plate tectonics.

Since coastline areas are among the most densely populated and developed land areas, humans spend a great deal of effort and money attempting to mediate the natural forces that make them so volitile. Coastline remediation has grown to become an important and expensive activity of government and private business, especially in areas where the economy is dependent on shoreline development. Not surprisingly, remediation has great significance as an environmental issue.

The *Earth Matters* CD focuses on the various aspects of natural forces effecting coasts. Sea level variations, plate tectonic activity, waves and currents, and shoreline types are all explored in the unit. The section on waves and currents includes animations depicting wave refraction, longshore drift, and rip currents. The unit concludes with a section on remediation that graphically shows the effects of five methods of coastal remediation.

Review

1.   What happens at the shoreline when the sea level lowers?

_____

2.   What happens at the shore line when the sea level rises?

_____

3.   During mountain building, what causes a coastline to prograde, or move outward into the sea?

_____

4.   Give an unintended effect of a jetty.

_____

_____

3. Wind

In many arid parts of the world, wind is the primary agent of change.

Learning Objects

- To be able to describe how wind patterns form

- To be able to list and describe the four principal types of dunes

- To be able to list three ways in which deserts may form

Introduction

Wind is the result of air moving from regions of high pressure to low pressure created by the differential heating of the earth's surface. At the equator, solar heating causes warm air to rise and creates a high-pressure zone. In the polar regions, the sun's rays strike the earth at a low angle, creating colder and drier air and a high-pressure system. The cold air moves toward the low-pressure area in the lower latitudes. The air from the equator moves toward the poles, then sinks as a cold dry air mass at about 30 degrees north and south, a location known as the horse latitudes.

As you work through this unit on *Earth Matters*, you will be able to see how winds form deserts, and within the deserts, sand dunes. It includes animations depicting dune formation in an introduction. The final section—desert formation—shows you how deserts form. On an island, you are to choose the direction of the prevailing winds as well as the placement of mountains. You will then see what sorts of deserts formation occurs as a result of your choices. By varying your choices, you should gain a deeper understanding of desert formation and the factors that determine it.

Review

1. What causes air to circulate from regions of high pressure to regions of low pressure?

    _____

2. In what direction does polar cold air flow?

    _____

3. In what way do blowout dunes differ in appearance from barchan dunes?

    _____

4. Glaciers

Glaciers occupy only a small part of the earth's surface.

Learning Objectives

- To be able to explain the mechanics of glacier formation and movement

- To be able to list and define the various types of glacier-created landforms

Introduction

*E*arth scientists believe that there have been several main ice ages, each of which lasted a few million years. The most recent ice age occurred mainly during the Pleistocene Epoch, which began about 2 million years ago and ended about 10,000 years ago. Of the many ice ages that have occurred, it is this one that scientists know the most about and usually mean by the term *the Ice Age.*

During glaciation periods, great continental ice sheets form, accumulating from about 8,000 to 10,000 feet thick. In the most recent ice age in North America, the main center was formed around Hudson Bay, from which it spread outward to flow in all directions. It covered most of North America to about the present valleys of the Missouri and Ohio rivers.

Some scientists believe that recurring changes in the shape of the earth's orbit around the sun cause cooling trends that promote the formation of huge ice sheets. Glaciers form when more snow falls in the winter than can melt in the summer. Snow forms as hexagonal ice crystals, which then metamorphose to form glacial ice. The crystals melt at the outer edge of the flake and refreeze toward the center to eventually form a firn grain of ice. After years of annual accumulation, the firn grains compact under the weight of annual accumulation. During the warmer months, snow melts and percolates through the firn pack and re-freezes, forming denser ice. When the ice reaches a sufficient thickness (30 to 40 m) that it can no longer support its own weight, it begins to flow under the forces of gravity.

This unit on *Earth Matters* will explain the relationship between the mass balance of a glacier and its movement, and it will also illustrate various landforms that glaciers leave behind in their wake.

Review

1.   What force causes glaciers to move?

_____

2.   What is till?

_____

3.   What happens when glacial ice reaches a thickness of 30-40m?

_____

5. Capstone for Surface Processes

An exercise in placing surface processes within the earth system context

The four exercises in this capstone unit will ask you to identify the locations of geologic processes in four settings—a beach, a desert, a high mountain stream, and a glacial area.

Correlation of Module to Freeman Textbooks

EG: *Environmental Geology* by Dorothy Merritts, Andrew De Wet, Kirsten Menking
UE: *Understanding Earth*, Second Edition, by Frank Press and Raymond Siever

Overall Module

EG: Coverage of this group of topic is not concentrated in any one area of the textbook.

UE: The coverage of many of the chapters inPart 2 (Chapters 11-17) correlates closely to that of this module.

1. Rivers

EG: Chapter 7: The Surface Water System, pp. 192-231

UE: Chapter 13: Rivers, pp. 316-343

2. Coastal Processes

EG: Chapter 10: The Oceans and Coastal System, pp. 290-324

UE: Chapter 17: The Oceans, pp. 418-455

3. Wind

EG: "Soil Erosion Hazards and Soil Conservation" in Ch. 6, pp. 179-182; "Arid Environments" in Ch. 12, pp. 373-374

UE: Chapter 14: Winds and Deserts, pp. 344-369

4. Glaciers

EG: "Indicators of Environmental Change" in Ch. 12, pp. 372-376; "The Little Ice Age" in Ch. 13, pp. 399-400

UE: Chapter 15: Glaciers, pp. 370-399

5. Capstone for Surface Processes

Both: This unit is an exercise related to the coverage of the other units in the module.

 See the web site for URL links to sites that relate to the topics in this module.

www.whfreeman.com/geology

Module Four
Groundwater

*T*he enormous reservoir of groundwater stored beneath the earth's surface equals about 22 percent of all the freshwater stored in lakes and rivers, glaciers and polar ice, and the atmosphere.

Groundwater forms as raindrops infiltrate soil and other unconsolidated surface materials, sinking even into cracks and crevices of bedrock. Beds that store and transmit groundwater in sufficient quantity to supply wells are called *aquifers*. The condition of aquifers is monitored to ensure that there is an adequate supply of freshwater to meet the needs of human consumption. Areas that are densely populated must be monitored carefully to avoid freshwater shortages.

The pollution of freshwater in industrialized countries is almost universal. For centuries, freshwater has been the cheapest method of disposing of waste, and industrial development has relied on it to lower the costs of manufacturing goods and providing services. Residential developments have also taken advantage of this apparently convenient and inexpensive waste disposal method. Historically, groundwater has been removed, contaminated, and then either put back into the ground where it spreads out in plumes and contaminates more groundwater or put into the coastal salt water in the hopes that the tides and currents would take it effortlessly out to sea.

The experience of the Boston metropolitan area is a typical example of how a large human population can pollute once pristine groundwater. The Pilgrims landed in Plymouth, Massachusetts, in 1620. They arrived and set up camp in an area that was already populated by the Wampanoag Indians. Only half of the Pilgrims who arrived in the new land survived the winter. Nevertheless, more settlers followed, and the area grew in population and commerce at a spectacular rate. Seventy-five miles away from the first settlement, Boston was founded in 1630. By 1990, the population of the Boston metropolitan area, comprising 92 cities and towns, was about 2.75 million people.

The new country began in an area of America that had bountiful freshwater supplies. Unfortunately, most of it has been polluted as development and industrialization has intensified. As New England was profiting in other ways from its factories and cities, it was consuming its freshwater resources at a significant rate. Today, the Massachusetts Water Resources Authority, which responsible for providing water and waste-disposal means to the 43 cities and towns in and around Boston, must reach all the way to Quabban Reservoir, located 65 miles away, to meet its needs for freshwater.

Freshwater in coastal zones is particularly vulnerable to contamination by saltwater intrusion, as examined in the first unit in the module. Geologists use Darcy's law and their knowledge of the porosity and permeability of rock materials to gauge the potential and actual water discharge and recharge rates in aquifers. The capstone allows CD-ROM users to change well digging sites and depths to make the most of the water availability in a particular area.

Units

1. Saltwater Intrusion

Five stages of coastal zone development as it affects rates of groundwater use and subsequent saltwater intrusion

2. Darcy's Law

The calculations used in determining the rate of groundwater flow through an aquifer

3. Porosity and Permeability

The porosity and permeability of gravel, sand, silt, sandstone, shale, limestone, basalt, volcanic tuff, slate, clay, granite, gneiss, fractured granite, glacial till, and sand and gravel

4. Capstone for Groundwater

An evaluation of groundwater supply problems and solutions

1. Saltwater Intrusion

Excessive drawdown in coastal zones creates an intrusion of saltwater into freshwater aquifers.

Learning Objectives

- To be able to identify the five stages of coastal zone development as that development affects the freshwater supply

- To be able to list three strategies that can prevent excessive saltwater intrusion

Introduction

In addition to the continual effort of keeping pollution away from their freshwater sources, people who live near the ocean may face another source of contamination—saltwater intrusion into their freshwater aquifer. This phenomenon occurs when freshwater is pumped out of the aquifer at a high rate in relation to recharge. Without human settlement the groundwater system of coastal areas would be in a state of dynamic equilibrium in which groundwater discharge and recharge would be in balance. Human population depends heavily on groundwater, so discharge inevitably outpaces recharge when human population is present in sufficient numbers. Left unchecked, this can lead to salt groundwater taking the place of fresh groundwater in coastal areas. A number of remediation tactics exist to combat saltwater intrusion.

This unit on *Earth Matters* examines the relationship between groundwater consumption and recharge in coastal zone areas. It graphically shows this phenomenon in five stages—from pre-settlement to actual intrusion to the introduction of countermeasures.

Review

1.   How many states in the United States have coastal frontage?

_____

2.   What is the primary source of freshwater aquifer recharge in coastal zone developments in stages 1 through 4?

_____

3.   What are the two primary strategies that are successful in lessening saltwater intrusion?

_____

_____

2. Darcy's Law

Permeability and the hydraulic gradient are important to the rate of groundwater flow.

Learning Objectives

- To be able to describe the numerical relationship between the type of material and the rate of groundwater flow through it

- To be able to define hydraulic conductivity, hydraulic gradient, porosity, and permeability

Introduction

Darcy's law quantifies the rate of groundwater flow through an aquifer. This rate is calculated using the following formula:

$$Q = K(a)(i)$$

where Q = discharge, cm^3/s

K = hydraulic conductivity, cm/day

a = cross-sectional area, m^2

i = hydraulic gradient, m/m

This important formula helps geologists predict loss and gain to groundwater supplies.

In this unit on *Earth Matters*, you will be able to test different materials for their ability to permit groundwater to flow through them. You are asked to set the variables of aquifer material, cross-sectional area, and hydraulic gradient in an electronic Darcy's law calculator. By manipulating the factors in the model you can get a stronger idea of the meaning behind and the relationship between the numerical variables.

Review

1.   The material type for an aquifer is sand. The cross-section area is 1 and the hydraulic gradient is 1/3. What is the rate of groundwater flow through the aquifer?

_____

2.   Through which material would groundwater flow faster, clay or gravel?

_____

3. Porosity and Permeability

Porosity and permeability vary widely among different rock materials.

Learning Objectives

- To be able to define porosity

- To be able to define permeability

- To be able to give several examples of materials with high and low porosity and permeability

Introduction

The two factors that govern the rate of groundwater discharge and recharge are the rock materials' porosity and permeability.

- The porosity of rock material, the percentage of its total volume that is taken up by pores, determines how much it can hold. Porosity depends on the size and shape of the grains and how they are packed together. The more loosely packed the particles, the greater the pore space between the grains will be.

- The permeability of rock material describes is ability to allow fluids to pass through it. A material therefore can be porous but not permeable.

This unit on *Earth Matters* explains porosity and permeability. It also quantifies and explains the porosity and permeability of 15 rock materials which are commonly found in groundwater systems.

Review

1. Gravels have excellent porosity and permeability. Why, then, do they make poor aquifer materials?

   _____

2. How does sand hold water?

   _____

3. The porosity and permeability of sandstone depends on what factors?

   _____

4. Capstone for Groundwater

An exercise in evaluating groundwater supply problems and solutions

The capstone exercise on this unit of *Earth Matters* allows you to experiment with different solutions to groundwater maintenance problems, using factors identified in the preceding units in this module. Using a "drag and drop" mechanism, you are prompted to decide where to drill wells and to pick the best location for an aquiclude in a groundwater system. As a result you should leave with an increased understanding of how aquifers work.

Correlation of Module to Freeman Textbooks

EG: *Environmental Geology* by Dorothy Merritts, Andrew De Wet, Kirsten Menking
UE: *Understanding Earth*, Second Edition, by Frank Press and Raymond Siever

Overall Module

EG: Chapter 8: The Groundwater System, pp. 232-259

UE: Chapter 12: The Hydrologic Cycle and Groundwater, pp. 286-315

1. Saltwater Intrusion

EG: "Intrusion of Salt Water," in Ch. 8, pp. 249-250

UE: "Balancing Recharge and Discharge" in Ch. 12, pp. 301-302

2. Darcy's Law

EG: "Wells" in Ch. 8, pp. 239-242; "Darcy's Law and the Flow of Water and Contaminants in Rocks and Sediments" in Ch. 8, pp. 241-242

UE: "The Speed of Groundwater Flows" in Ch. 12, pp. 303-304

3. Porosity and Permeability

EG: "Porosity and Groundwater Storage" in Ch. 8, pp. 235-237; "Permeability and Groundwater Storage" in Ch. 8, pp. 237-238

UE: "How Water Flows Through Soil and Rock," pp. 296-298

4. Capstone for Groundwater

Both: This unit is an exercise related to the coverage of the other units in the module.

See the web site for URL links to sites that relate to the topics in this module.
www.whfreeman.com/geology

Module Five
Natural Hazards

*T*he chief difference between purely physical geology and environmental geology is that environmental geology focuses on the geological effects of the environment in which humans live. The emphasis on the consequences of geological events and factors on humans is what distinguishes the study from its older cousin. Volcanoes, floods, earthquakes, and mass wasting are all events that have rich and interesting stories based in physical geology. Some of this is covered in other modules in *Earth Matters*. This module concentrates on their environmental and human aspects.

Volcanoes, floods, earthquakes, and mass wasting devastate landscape, spread pollution, destroy property, and take human life—often in enormous numbers at one time. The units in this module emphasize the prevention of human deaths and injuries through the development of early warning systems and (whenever possible) preventative measures. Thus, the units evaluate the depth of our understanding of these catastrophes as well as the accuracy and reliability of methods of predicting the timing, duration, and strength of natural disasters.

Units

1. Volcanoes

The facts on five significant volcanoes that have occurred in the past, the hazards commonly associated with cinder cone, composite, and shield volcanoes, and the accuracy and value of predicting volcanic eruptions

2. Flooding

The relationship between flooding possibility and floodplain development

3. Earthquakes

Plate tectonic activity as a cause of earthquakes, types of damage, facts on some major earthquakes that have occurred, predicting earthquakes, and setting sound public policies to prevent deaths, injuries, and property damage

4. Mass Wasting

An analysis of the agents that contribute to mass wasting: slope, materials, rainfall, vegetation, and human impact

1. Volcanoes

*Public policy and safety are important factors in volcanic areas because
volcanoes pose a significant risk to human life.*

Learning Objectives

- To be able to describe some of the major volcanoes that have occurred in the past

- To be able to list some of the hazards of volcanic eruptions

- To be able to evaluate the usefulness of current methods of predicting volcanic activity

Introduction

Volcanoes are among the most violent and destructive—as well as probably the most dramatic—geological events on the earth. With relative frequency, large numbers of deaths have been lost on the occasions of large volcanic eruptions as they have occurred over the course of recorded history. Research on volcanoes, of necessity, focuses on life-saving measures for people living in their vicinity. Here geologists must grapple with the psychology of human beings: If they call for an evacuation and it turns out that it was not necessary, the next time they try to evacuate an area, the inhabitants may refuse to go. Thus, accuracy is critical—and unfortunately sometimes elusive. The good news is that we are making steady progress in estimating the timing of volcanic eruptions.

The eruption of Mount St. Helens in 1980 is one of the best known partly because it was so well documented and partly because its eruption was unusual. Geologists from the U.S. Geological Survey were very successful in predicting the eruption, which accounts for the very low loss of life (57 people). The eruption was the result of pressure buildup, which culminated in the volcano's blowing up. The explosion blew down over 600 acres of trees and produced a mudflow. An area of 500 km² was devastated by the blast, and ash fallout was experienced as far away as Denver. Only those people who refused to be evacuated died.

Volcanic eruptions are more easily predicted than earthquakes; however, correct prediction is more vital in order to save lives because the forces of volcanic eruptions cannot be mitigated by engineering to the extent that those of earthquakes can. Volcanoes have caused more death and destruction throughout history than earthquakes.

This unit on *Earth Matters* describes various hazards of volcanoes, gives a brief description of some noteworthy volcanoes of the past, and discusses the how the future may benefit from increasingly sophisticated and accurate methods of predicting the occurrence of volcanic eruptions.

Review

1.   Of the three volcanic types represented on the CD, which one poses the greatest threat of producing lahars?

_____

2.   When did Mount St. Helens last erupt?

_____

3.   Which volcanic danger did the appearance of blue flames at Mount St. Helens in March 1980 indicate to scientists?

_____

2. Flooding

The storm type and runoff conditions determine the extent and magnitude of flooding.

Learning Objectives

- To be able to explain why people maintain communities in floodplains

- To be able to list some climatic causes of flooding

- To be able to list some prevention strategies for floodplains

Introduction

*F*loods are the most widespread of catastrophic hazards, and they take thousands of lives and cause billions of dollars in property damage annually. In spite of this, humans will always be attracted to rivers and their floodplains for amenities such as freshwater, food, and easy access to transportation. Also, river floodplains provide excellent grazing and fertile soil for raising crops.

It is obviously no secret to those living near rivers that flood will periodically take place. This is why, increasingly, people have tried to protect themselves from flood waters with a variety of artificial structures. However, little if any success in preventing the larger floods has been realized—a fact that makes it into the news with great frequency.

After a flood, people will move back into flood-prone areas knowing that they probably will be flooded out again in the future. Thus, monitoring storms and water levels in populated floodplains is crucial to preventing human injury and death.

*T*he CD looks at the exacerbating role of urbanization in floods and flooding. You will learn and see how the products of urbanization contribute to the severity and frequency of major floods One of the interactive functions on this unit on *Earth Matters* will allow you to increase floodplain development through various levels to visualize how the increases affect the risk of flooding.

Review

1.   What is the definition of *flooding*?

_____

2.   What four factors most strongly affect the rate of surface runoff?

_____

_____

3.   List some of the impermeable surfaces found in urbanized areas.

_____

3. Earthquakes

Earthquakes are not as predictable as volcanic eruptions,
but their disastrous effects can still be mitigated by astute public policy.

Learning Objectives

- To be able to describe the types of hazards and damage that accompany major earthquakes

- To be able to list and describe the three types of earthquake effects that cause damage: liquefaction, swaying, and resonance

- To be able to list and evaluate the primary means available for predicting earthquakes: monitoring foreshocks, monitoring changes in elevation in the ground, monitoring the electrical resistivity of rocks, monitoring animal behavior, monitoring groundwater variations, and monitoring seismographs

- To be to explain the need for public policies that prevent or ameliorate the hazards of major earthquakes such as the consistent enforcement of earthquake-resistant building practices

Introduction

*P*late-tectonic activity is the major, although not only, cause of earthquakes. Earthquakes of low to moderate levels are common at divergent plate boundaries. The tension and extension that occur at such boundaries gives rise to normal faults. Although earthquake magnitudes tend to be lower on normal faults at divergent boundaries, there are several instances in the historical record in which major earthquakes occurred along divergent boundaries.

Many of the earthquakes that occur each year are located near convergent plate boundaries. The earthquakes that occur in Japan and the Philippines and those that occur in South and Central America are all related to the tectonic activity associated with convergent boundaries. The high compressive stresses at convergent boundaries lead to reverse and thrust faults and major earthquakes along with them. Earthquakes are one of the ways in which stress is released along a fault.

Transform or oblique slip boundaries are relatively rare, but they provide an important setting for earthquake activity. Two of the most seismically active areas in the world occur at transform boundaries: the San Andreas and the North Anatolian fault systems. The continuous sliding of one plate horizontally past another gives rise to long and complicated fault systems in which complex earthquake activity occurs.

*T*his unit on *Earth Matters* examines earthquakes from several perspectives as they affect human life on the earth. Types of damage are described and demonstrated in videos. The earth's plate boundaries are visually identified by type. A historical review looks at famous earthquakes and lists estimates of human lives lost to them. Several interesting but ultimately unsatisfactory methods of predicting earthquakes are described and the sticky question of the public policy ramifications of trying to predict earthquakes is discussed.

Review

1. What type of earthquake damage effect probably poses the most danger to man-made buildings?

_____

2. Earthquake activity at transform boundaries is relatively rare because this type of tectonic activity is rare, but there are two such seismically active areas in the world. Where are these two areas?

_____

3. What is the name of the term for changes in the elevation of ground near earthquakes due to dialation of deformed rock?

_____

4. Mass Wasting

More of the earth is affected by mass wasting than by earthquakes and volcanoes combined.

Learning Objectives

- To be able to evaluate the risk factors associated with mass wasting: slope, materials, rainfall, vegetation, and human impact

- To be able to describe some strategies for preventing mass wasting

Introduction

*M*ass wasting is the downslope movement of soil, sediment, and rock. It causes more damage in the United States each year than earthquakes, volcanoes, tornadoes, and hurricanes combined. Mass wasting occurs when the force of gravity exceeds the strength, or the resistance to deformation, of the slope materials. Such movements are triggered by earthquakes, floods, or other geological events. The materials then move down the slope, either slowly or very slowly or very suddenly, sometimes catastrophically.

Mass wasting can be destructive and deadly. One of the worst cases occurred in the Andes Mountains of Colombia in 1985 when more than 20,000 people lost their lives in a giant mudflow of unconsolidated volcanic ash. Geologists evaluate several risk factors in predicting these events.

*I*n this unit on *Earth Matters*, you will learn about mass wasting through the model of a cross section of a building site that can be manipulated. At this model site, variable conditions associated with mass wasting—slope, ground and bedrock materials, rainfall, vegetation, and human impact—can be investigated and thus more fully understood through a mechanism that lets you set and reset the conditions at different levels.

Review

1.   What type of "materials" situation is the most stable?

_____

2.   How does vegetation offset the risk of mass wasting?

_____

_____

3.   When soil materials predominate, a slow form of mass wasting is a potential danger. What is the term for this slow form of mass wasting?

_____

Correlation of Module to Freeman Textbooks

EG: *Environmental Geology* by Dorothy Merritts, Andrew De Wet, Kirsten Menking
UE: *Understanding Earth*, Second Edition, by Frank Press and Raymond Siever

Overall Module

EG: The coverage of the materials in this unit is not concentrated in any one area of the Merritts textbook.

UE: The *Understanding Earth* coverage of the materials in this unit is not concentrated in any one area.

1. Volcanoes

EG: Chapter 2 Introduction, pp. 30-31; "Predicting the Eruption of Mount St. Helens" in Ch. 3, pp. 67-68; "Volcanic Eruptions" in Ch. 5, pp. 138-146

UE: "Volcanism and Human Affairs," pp. 125-131

2. Flooding

EG: Chapter 7 Introduction, pp. 194-195; "The Hazards of Flooding" in Ch. 7, pp. 206-216

UE: "Stream Valleys, Channels, and Floodplains," pp. 323-326; "The Development of Cities in Floodplains," p. 327

3. Earthquakes

EG: "Earthquakes" in Ch. 5, pp. 146-154

UE: "Earthquake Destructiveness," pp. 472-481

4. Mass Wasting

EG: "Mass Movement Hazards and Their Mitiagation" in Ch. 6, pp. 182-188

UE: Chapter 11: Mass Wasting, pp. 264-285

✓ *See the web site for URL links to sites that relate to the topics in this module.*
www.whfreeman.com/geology

Module Six
Waste Disposal

Waste is generated as a by-product of almost every conceivable human endeavor. In ages past, most of the waste produced was organic, and given sufficient time and a reasonably sized disposal area in proportion to the amount of waste to be broken down, it was usually reabsorbed safely into the biological environment without ever causing harm to human health. In contrast, the twentieth century has seen a virtual explosion of technological and chemical inventions that have immeasurably improved living conditions for human beings, but at a great cost that was at first largely overlooked. The unanticipated result of economic expansion has been a growing problem of non-biodegradable waste disposal.

Different chemicals react differently, either alone or in combination with other chemicals, in various settings. So the waste disposal problem must be resolved with a twofold approach: by the type of material to be disposed of and the medium into which the waste is to be disposed. This issue, then, is extremely complex.

The first two units in this module examine the types of waste that must be somehow disposed of and the possible places where it may be disposed of and safely confined. The second unit and the capstone unit are both exercise-oriented, and both deal with the key environmental topic of groundwater.

Units

1. Waste Types

The sources of waste: industrial and commercial, domestic, and government, as well as the types of waste products: solid, toxic, and radioactive

2. Site Selection

Assessing the risk of spreading contamination from a radioactive waste storage facility allowing for the following factors: subsurface materials, plate-tectonic activity, topography, human behavior, and ground and surface water

3. Capstone for Waste Disposal

An exercise in finding the pollution source and choosing a remediation method

1. Waste Types

Waste types and sources are important factors in developing national waste disposal strategies.

Learning Objectives

- To be able to name the leading public- and private-sector producers of solid waste

- To be able to name the leading public- and private-sector producers of toxic waste

- To be able to name the leading public- and private-sector producers of radioactive waste

Introduction

*O*ne of the best-known cases of chemical pollution in American history occurred in the Love Canal neighborhood of Niagara Falls. By the 1970s, chemical wastes buried in a local disposal site had percolated underground and traveled quite a distance to a residential area. There they rose to the surface and contaminated drinking water supplies and the yards of local homeowners, threatening the health of the area's residents. As a result of several class-action lawsuits, both the New York state government and the federal government eventually provided financial aid to help move the families from Love Canal to other areas. This case brought to the public's attention the need for formal and strict controls on the disposal of certain types of waste.

The Resource Conservation and Recovery Act (RCRA) of 1976 and its 1984 amendments were legislative responses to the increasing public concern over the proliferation of chemical wastes in the environment. The RCRA lists the properties that make a waste hazardous. The law declares a waste to be hazardous if it corrodes (or wears away) other materials; explodes; is easily ignited; reacts strongly with water; is unstable to heat or shock; or is poisonous. Poisonous wastes are commonly called *toxic wastes*. The law also covers radioactive wastes that occur as by-products of mining operations or as a result of the use of radioactive substances in medical procedures. Radioactive wastes from commercial power plants and other nonmilitary sources are controlled by the U.S. Nuclear Regulatory Commission, and radioactive wastes from military sources are controlled by the U.S. Department of Energy.

The Comprehensive Environmental Response, Compensation, and Liability Act of 1980, also called "the Superfund," provided $1.6 billion to clean up unsafe dump sites. A 1986 act reauthorized the Superfund and provided an additional $9 billion. Today there are literally hundreds of sites on the Superfund list. It has been estimated that it would cost over $100 billion to restore all of them.

In this unit on *Earth Matters*, you will see in graphic representations a summary of the major types and sources of problematic waste in the United States. Some of the information contained in this unit may surprise you—such as the fact that construction accounts for more solid waste than residential sources. The section on radioactive wastes shows that while nuclear power plants have a good safety record in the U.S., the longevity of toxicity of the waste they produce raises great environmental concern.

Review

1. What is the second leading source of commercial solid waste in the U.S.?

_____

2. What sorts of hazardous wastes are generated by farming activities

_____

3. Why is radioactive waste more difficult to dispose of than other types of waste?

_____

_____

2. Site Selection

Proper site selection is the key to safe waste disposal.

Learning Objectives

- To be able to assess the risk of pollutants' leaking and spreading in a particular geographical area based on the following factors: subsurface materials, plate-tectonic activity, topography, human behavior, and ground and surface water

- To be able to predict the consequences of contamination spreading from a particular waste disposal site

Introduction

*D*isposal of radioactive waste is a difficult process because the life cycle of the waste is extremely long and the effects of a breach of the disposal site are catastrophic to humans. Another factor that complicates the task of site selection is the fact that there is really no such thing as a perfect site. At its very core, site selection involves weighing positive factors against negative ones.

*T*he site shown in the geological map and profile in this unit on *Earth Matters* contains the Hanford radioactive waste storage facility. This site has been used by the U.S. Department of Energy and the U.S. Department of Defense since the 1940s and contains tons of radioactive waste. In the 1940s, when less was known about the hazards of radioactive waste and little study of geologic factors of disposal sites had been done, Hanford—with its relative remoteness and other favorable factors—might have seemed like a suitable site for selection. As you will see, however, there are a number of difficult or impossible to solve geologic drawbacks to the site that make Hanford seem like a relatively poor choice. When you are done with this unit, you might want to consider what sort of geologic conditions would constitute a choice for radioactive site selection.

Review

1.   What hazard would an earthquake pose to a radioactive waste storage facility?

_____

2.   What hazard would flooding pose to a radioactive waste storage facility?

_____

_____

3.   What is a favorable factor for radioactive waste storage relating to the groundwater at Hanford?

_____

4.   What possible danger would there be in moving the radioactive storage bunkers at Hanford above gound?

_____

3. Capstone for Waste Disposal

An exercise in finding the pollution source and choosing the remediation method.

*I*n this exercise you will direct the analysis of an area in which the groundwater has been polluted. Your task is to locate the source of pollution and determine whether the source is a point source or an area source. You will be able to examine each of the six subdivisions of the map provided on the CD-ROM and take well water samples and determine the nature of each factory or other potential source. The water wells are represented by red dots, and the potential sources are blue.

A consultant will be available to you to help you determine the source and the strategy to clean up the site. You have a budget of $50 million. The costs of the services of the consultant, the well water analyses, and the cleanup will be charged against your budget.

After you have chosen a means of remediation, you will be told the outcome of your strategy. As is usually the case in the real world of waste disposal, there are no clear cut "right" or "wrong" answers in this exercise. Every "solution" has consequences, and thus, the trade-off nature of environmental decision making is illustrated.

Correlation of Module to Freeman Textbooks

EG: *Environmental Geology* by Dorothy Merritts, Andrew De Wet, Kirsten Menking
UE: *Understanding Earth*, Second Edition, by Frank Press and Raymond Siever

Overall Module (all Units)

EG: Chapter 3 Introduction, p. 64; "Predicting the Geologic Stability of Yucca Mountain" in Ch. 3, pp. 82-83; "Butte, Montana—From Boom Town to Superfund Site" in Ch. 5, pp. 132-133; "Surface Resources and Protection" in Ch. 7, pp. 220-228; "Groundwater Pollution and Its Cleanup" in Ch. 8, pp. 250-256; Chapter 10 Introduction, pp. 292-293; "Ocean Pollution" in Ch. 10, pp. 316-321

UE: "Water Quality" in Ch. 12, pp. 308-311. Most of the *Understanding Earth* coverage of waste disposal relates to radioactive waste and is found in "Radioactive Waste Disposal," pp. 574-575; and "Nuclear Energy Hazards," pp. 575-576. Also see "The Oceans as a Deep Waste Repository," p. 450.

See the web site for URL links to sites that relate to the topics in this module.
www.whfreeman.com/geology

Module Seven
Resources and Sustainability

*T*he environmental branch of geology has many of its roots in the studies of the topics in this module. The earth contains various strategic resources—both minerals and fossil fuels—that are crucial to the way of life that many of us have grown used to enjoying. The recognition that there are limits to the supplies of these minerals and fossil fuels, and thus that we must be careful if we are to sustain the lifestyle to which we have grown accustomed, was an important step in the growth of environmental geology.

The first unit in this module profiles the fuels and strategic materials the industrialized world depends upon to maintain its way of life. The second unit reviews the status of the strategic materials supplies in 15 developed countries, including the United States. Each country's ability to meet its own needs is examined individually.

The third unit examines plausible events that could occur in response to an oil and natural gas shortage in the United States. In this unit, you view the situation from the Oval Office which you occupy as the President of the United States. You are given contingency plans by many advisors. The situation is critical, and your decisions will have enormous consequences.

The fourth unit in this module—the capstone—casts you as a "wildcat" oil driller, trying to discover a source of petroleum. You have access to a map and to a polite but aloof consultant. Your task in this unit is to find petroleum before your money runs out.

Units

1. Strategic Materials

The various types and availability of energy resources and strategic mineral and metallic materials

2. Resources Politics

The production and consumption of strategic materials and fossil fuels in 15 countries: Bolivia, Brazil, China, Germany, Great Britain, Mexico, Peru, Russia, Saudi Arabia, South Africa, Turkey, the United States, Zaire, Zambia, and Zimbabwe

3. Public Policy

The ramifications of government actions to protect strategic materials supplies

4. Capstone

An exercise in exploring for oil and natural gas

1. Strategic Materials

Strategic materials are the fuel for industrial and economic growth.

Learning Objectives

- To learn about the availability and use of strategic materials

- To learn about the availability and use of fossil fuels

Introduction

*F*ossil fuels are central to the stability of the world's economy. They supply most of the world's energy needs—petroleum, coal, and natural gas—as well as strategic materials used in manufacturing manganese, bauxite, platinum, cobalt, chromium, zinc, and iron. In fact, people are so dependent on these resources that any interference in the steady rate in which they are mined and made available to industries and governments threatens their livelihoods, and therefore their survival.

*T*his unit on *Earth Matters* looks at the makeup, production, and consumption each of the principal resources upon which the world's industries depend. It also lists the most important source countries for many of these materials. All of this information will be useful in your understanding of the other units of this module.

Review

1. According to the U.S. Geological Survey, how long will the world's petroleum reserves last?

    _____

2. According to the U.S. Geological Survey, how long will the world's natural gas reserves last?

    _____

3. What industries use the most chromium?

    _____

4. Name a strategic material that South Africa produces in major quantities?

    _____

5. What strategic material is found in the Lake Superior region of the U.S. and Canada?

    _____

2. Resources Politics

Resources distribution is an international issue of such importance that it can lead to war.

Learning Objectives

- To be able to describe various countries from the standpoint of their production and consumption of strategic materials

- To be able to describe the unique position of the United States in the world with regard to its vast deposits and enormous consumption of strategic materials

Introduction

The industrialized world depends on uninterrupted supplies of raw materials and cheap energy to sustain its complex, highly specialized existence. Unfortunately, the raw materials and fossil fuels upon which the world relies are finite in quantity and located in some countries but not others, even though the need for them is universal among all developed countries. To survive, countries must import some of the materials upon which they are the most dependent from other countries, which means that international relations have a direct bearing on the quality of modern life.

The developed world was quite stunned in the 1970s when the Arab nations, among them Saudi Arabia, formed a trade group called OPEC, which over a short period of time nearly doubled the per-barrel prices of the petroleum they exported to other countries. The Middle East has vast supplies of petroleum and natural gas, with as much as 25 percent of the world's petroleum in Saudi Arabia alone.

Caught by surprise, the governments trading with OPEC took many actions to avert economic disaster. The short-term measures involved asking people to conserve energy, to drive slower, and to turn down their thermostats, among others. The long-term measures involved developing strategies for lessening each country's dependence on the OPEC supply of fossil fuels. Some countries invested in petroleum mining development programs that were to be maintained wholly within their own borders. Others experimented with alternative fuel sources such as wind, water, and solar energy. Still others, lacking internal sources of fossil fuel supplies, opened up trade with as many other fuel producing nations as possible so that they would not be too dependent on any one country.

As you proceed through this unit *Earth Matters*, you may be surprised at which countries are most and least dependent on other countries for their fuel and strategic materials supplies. You will note that there is little relation among factors of population, consumption, and production. You will also see that many small countries are among the most important produces of strategic materials.

Review

1.   What does the term *net producer* mean?

_____

2.   What would the effect be on the rest of the world if the United States were removed as a supplier and consumer?

_____

_____

3.   What percentage of the world's fossil fuels does the U.S. produce?

_____

4.   Does Germany consume more energy than Great Britain?

_____

3. Public Policy

Public policy and science often clash.

Learning Objectives

- To be able to analyze U.S. short-term and long-term public policy for protecting strategic materials imports upon which the country depends

- To be able to analyze the effect on public opinion of actions taken by government to protect foreign sources of strategic materials

Introduction

This unit is an exercise using a hypothetical geopolitical situation involving the real life interdependence of nations with regard to strategic materials. Among other points, it is designed to again show you that serious environmental problems defy easy solutions.

In the exercise on this unit, a purely hypothetical crisis is described involving the United States and its long-time ally, the oil-rich nation of Saudi Arabia. Saudi Arabia has been chosen for this exercise only because it well illustrates the dependence of the U.S. on various countries that produce strategic materials. A great many of the circumstances described in the exercise would be similar if the U.S. were to lose access to the strategic materials of many nations other than Saudi Arabia.

In this unit you are asked to put yourself in the position of the President of the United States when faced with a serious resource crisis. Saudi Arabia, a heretofore stalwart supplier of oil for the U.S., has decided to stop exporting oil. Saudi Arabia accounts for almost a quarter of all the oil imported by the United States and the other industrialized nations of the world. In the mid-1970s a similar reduction in the availability of oil led to a major recession in the U.S., a stagnated economy, high inflation, and a decrease in the sales of domestic automobiles. As President, you will be given advice from many congressional and cabinet agencies. You will explore the possible results of executing the contingency plans suggested to you. Each option you consider will have benefits *and* drawbacks. The unit illustrates that a challenge of public policy in environmental matters is the trading off of these factors to find just the right balance of benefits and drawbacks.

Review

1.   What percentage of current United States energy needs is met with petroleum and natural gas?

_____

2.   How did the public respond to the rationing of fuel during World War II?

_____

_____

3.   What percentage of the U.S. population uses public transportation and carpooling?

_____

4.   What is currently the most expensive form of energy in the U.S. in terms of cost per kilowatt hour?

_____

4. Capstone for
Resources and Sustainability

Exploration for resources is complex and expensive.

*I*n past decades, fossil fuels tended to dominate geological studies in schools since most people going on to careers in geology were going into the petroleum industry. At the present time, far more geology students are going on to environmental jobs (and very often, jobs relating to groundwater issues) than are going into the oil industry. Petroleum studies remain extremely important to geology in general, as well as to an understanding of environmental issues, however.

*I*n the unit on *Earth Matters*, you will be asked to search for oil and gas in a particular map location. You must be prudent in your exploration. Before you drill, you will need to take core samples from each section of the map in order to understand the subsurface geology. A consultant is available to provide you with advice and assistance. Although it costs more money to obtain core samples and consultant advice, it is less expensive to explore and drill intelligently than to drill blindly.

Correlation of Module to Freeman Textbooks

EG: *Environmental Geology* by Dorothy Merritts, Andrew De Wet, Kirsten Menking
UE: *Understanding Earth*, Second Edition, by Frank Press and Raymond Siever

Overall Module (all Units)

EG: "Resources and Sustainable Development" in Ch. 1, pp. 17-19"Earth's Energy System" in Ch. 2, pp. 47-50; "Rock and Mineral Resources" in Ch. 5, pp. 124-138; Chapter 11: Energy and the Environment, 326-361

UE: Part 4 (Chapters 22-24) "Conserving Earth's Bounty" covers the same general topic as this module—especially Chapter 22: Energy Resources from the Earth, pp. 560-583 and Chapter 23: Mineral Resources from the Earth, pp. 584-609

✓ ***See the web site for URL links to sites that relate to the topics in this module.***
www.whfreeman.com/geology

Module Eight
Field Trips

By its very nature, geology has always been a science where there is great potential for learning in the field. Unfortunately, the teaching of geology at the introductory level is almost always confined to the indoors. While there can never be a full substitute for the learning that takes place outside, new computer technology is making great strides in simulating the real with the "virtual."

Apple Computer's QuickTime Virtual Reality (QTVR)™ software is a leading tool in this technological and educational push. QTVR enables users to navigate freely (back/forth, up/down, as well as inward/outward) through panoramic vistas of remote sites that have been created through the "stitching" together of multiple images. "Hotspots" within these panoramas mark spots where further information—more detailed shots, specific data—can be revealed by clicking on the computer.

Earth Matters contains QTVR field trips of three sites—all places to which instructors of environmental geology might like to take their students in person if that were possible. The first field trip is a tour of five areas at a modern landfill in southern Indiana. The second examines three areas of active faults near and literally cutting right through the city of Hayward, California. The third field trip shows two areas in a part of California's Marin County where dramatic evidence of plate-tectonic activity is visible. You can use the rock evidence to formulate a hypothesis of the geological events that may have occurred there.

Within each area of each filed trip you may navigate freely in a wide (often 360 degree) vista. As your cursor move over specific points, called "hotspots," the computer screen signifies that your cursor is over a point of particular geologic interest. Click on these hotspots to reveal information specific to them. You also have the option of letting the computer automatically find the hotspots by clicking on the hotspot buttons within each area.

Units

1. Landfills

A virtual guided tour of a modern trash-handling facility in Monroe County in Southern Indiana gives you an opportunity to learn about the construction, operation, and monitoring of a sanitary landfill

2. Faults

A study of the fault-prone Hayward, California area to assess past and present earthquake activity, to gauge the risk for future earthquake occurrences, and to recommend strategies to protect human health and property

3. Tectonics

Exploring the rock materials of a site north of San Francisco with suspected plate-tectonic activity to determine the nature of the activity from the rock materials on the site

1. Landfills

*Maintaining and monitoring landfills are important
in protecting a community's groundwater from contamination.*

Learning Objectives

- To be able to describe the operations required to maintain a modern-day landfill

- To be able to describe the actions communities must take in operating their landfills to protect their groundwater supply

Introduction

While the local "dump" may not seem like a pleasant place, nor does it usually seem like a potential menace to human health. However, depending on where it is built and how carefully it is maintained and monitored, it can actually pose a grave threat to human beings. Toxic wastes can be deposited at a landfill by unscrupulous or unknowing people where the wastes can seep into the groundwater supply and poison inhabitants in the area.

Many communities in the United States are presently overwhelmed by the requirements of trash handling. Over the last decade, public education about the nature of toxic wastes is now a large part of the trash collection and storage jobs in medium and large cities and small towns alike. Household trash is as apt to be dangerous in a landfill as small- and medium-sized business trash.

To avert a potential disaster, most cities and towns are aggressively pursuing the use of recycling and hazardous-waste-disposal days when residents can bring their unused nail polish, household insecticides, and paint thinner to a specially designated drop-off point for proper disposal. Public education efforts also extend to businesses and institutions to inform management of the risks posed by ordinary-seeming refuse. However, much remains to be done to ensure a safe drinking water supply in this country.

As you explore the Monroe County Landfill in Southern Indiana on *Earth Matters*, you will see firsthand the rather imprecise and unscientific means available to maintain and monitor the waste generated every day in the United States. It is divided into four areas within the landfill:

1. This scene at a working landfill illustrates how the waste material is buried with soil taken from a nearby "borrow pit." A bulldozer is used to move and localize the waste deposited each day.

2. A critical concerns in a sanitary landfill is the monitoring and remediation of the leachate—contaminated water that percolates through the landfill, and concentrates some of the toxic ingredients in the waste. Leachate is pumped into a collection pool and remediated. You will have the opportunity to view the chemical content of the leachate and the water from surrounding water wells. A comparison of the chemical content of each lets you determine if the leachate is leaking into the local water table.

3. In this scene you will observe the creation of a new sanitary landfill. Because the site is under construction, you will have the opportunity to see a cross section and some of the important parts of the landfill.

4. In this scene you will observe the creation of a new sanitary landfill. One of the features of this scene is the leachate collection and remediation.

Review

1. What purpose does clay serve in a landfill?

_____

_____

2. What does the leachate monitoring well do with excess leachate?

_____

3. What is the final stage of landfill development?

_____

2. Faults

Touring fault zones is instructive in identifying and preventing earthquake damage.

Learning Objectives

- To be able to recognize fault damage in an outside area

- To be able to recognize earthquake-caused structural damage in a building

Introduction

As we have seen in earlier modules, predicting earthquake activity is not always possible, even with expert guidance and the latest technology. Therefore, in earthquake-prone areas, scientists and engineers focus their energies on construction practices that mitigate or prevent, most importantly, the injury or loss of human life or, secondarily, damage to property.

On this field trip you will be in and around Hayward, California, a densely-populated and highly fault-prone area to the south of San Francisco. The Hayward Fault cuts directly through the middle of town. As you will see in one area, it runs directly under the former city hall, which has been closed and padlocked shut to reduce the potential danger to humans.

The only stretch of the fault near Hayward that has not been highly developed by humans is a few hundred yards of land bordering on a cemetery, which you will see in another area. The other area is outside of Hayward, heading west toward Half Moon Bay. From the vantage point of a highway rest stop poised over an expansive valley, it allows you to explore a stretch of the San Andreas Fault—probably the most studied and most notorious fault in the world.

Compared to its known history, the Hayward Fault has been extremely quiescent in the 20th century. This leads many experts to the belief that major activity on this fault will probably occur relatively soon. The San Andreas Fault is also considered a tremendous ongoing risk to the great population in the highly-developed San Francisco Bay area.

With your camera's eye view, you will be able to assess earthquake risk and try out loss-prevention strategies.

Here is a description of the three areas within this field trip:

1. The Hayward fault is shown in this scene in a rural setting. Evidence of the presence of the fault can be observed in several areas of this sloped terrain.

2. The San Andreas fault system runs through the reservoir shown in this scene. You will notice that the landscape is dramatically altered by the fault. Differential weathering in the fault zone results in the creation of narrow valleys in the landscape along the fault. The granulation of rock is also illustrated.

3. When faults are present in urban areas the effects and signs of deformation. Some of the evidence for the presence of active faults in urban settings include the displacement of sidewalks, streets and structures, fracture patterns in structures and roadways, and variations in the amount and nature of vegetation.

Review

1.	What is the technical term for the cracks in the parking lot at the former town hall?

_____

2.	Why is there apt to be lush vegetation in an active fault zone?

_____

_____

3.	Why do modular structures withstand earthquakes better than other types of structures?

_____

_____

3. Tectonics

Rock material can be examined as evidence in
determining the nature of tectonic activity in a particular place.

Learning Objectives

• To be able to identify a type of tectonic activity from examining the rock material in a particular area

• To be able to explain seemingly contradictory rock findings in a particular area

Introduction

Geologists use rock material samples to assess the history of a particular place. Using the available historical evidence found in rocks, geologists are able to make intelligent observations about what has happened and predictions about what might be expected to happen in the future. On this field trip, you will be able to evaluate the plate-tectonic events that likely occurred based on the rock materials you find on the site. The site you will visit in this field trip is one that is particularly rich in rock evidence.

This field trip examines one of the important types of plate boundaries, convergent boundaries in which an oceanic plate in subducted under a continental plate. One interesting feature of this site is that it is now a oblique slip boundary that evolved from a convergent boundary. The features you will see in this trip are the result of the convergent period.

Just across the Golden Gate Bridge from San Francisco there is a park in an undeveloped area containing a former military installation. Its rugged beauty and the bounty of evidence relating the dramatic geologic history of the area make it a geologist's dream. Anyone with an interest in geology should make a point of visiting the site in his or her lifetime. As a substitute to actually being there and learning about the geologic wonder of the place firsthand, this virtual field trip is offered.

The field trip shows two areas in the park:

1. In areas of tectonic activity, especially near active plate margins, very different rock assemblages representing different environments are mixed together. In this scene you will see a fault that forms the boundary between, oceanic crustal materials, terrestrial rocks and melange rock formed when the two plates converge.

2. This scene represents one of the classic examples of a tectonic boundary in which plates converge and produce unusual features such as the mixing of oceanic and terrestrial rocks at the boundary.

Review

1. What type of tectonic activity would the presence of hydrothermal deposits suggest?

_____

_____

2. What makes the melange rock green?

_____

3. How are pillow basalts formed?

_____

_____

Correlation of Module to Freeman Textbooks

EG: *Environmental Geology* by Dorothy Merritts, Andrew De Wet, Kirsten Menking
UE: *Understanding Earth*, Second Edition, by Frank Press and Raymond Siever

Overall Module

EG: The coverage of the materials in this unit is not concentrated in any one area of the Merritts textbook.

UE: The *Understanding Earth* coverage of the materials in this unit is not concentrated in any one area.

1. Landfills

EG: "Surface Resources and Protection" in Ch. 7, pp. 220-228; "Groundwater Pollution and Its Cleanup" in Ch. 8, pp. 250-256

UE: The closest coverage in *Understanding Earth* can be found in "Water Quality" in Ch. 12, pp. 308-311 and the waste disposal coverage of Chapter 22.

2. Faults

EG: "Earthquakes" in Ch. 5, pp. 146-154

UE: "How Rocks Become Formed," 247-248; "How Rock Fracture," pp. 255-258; "Unraveling Geological History," 258-260; "What Is an Earthquake?" pp. 460-462

3. Tectonics

EG: "Plate Tectonics and the Rock Cycle" in Ch. 4, pp. 102-110

UE: "Plate Tectonics: A Unifying Theory for Geological Sciences," pp. 14-20; " "The Big Picture: Earthquakes and Plate Tectonics,: pp. 470-2; Chapter 20: Plate Tectonics, pp. 504-535

✓ *See the web site for URL links to sites that relate to the topics in this module.*
www.whfreeman.com/geology

Glossary

ablation   The annual amount of ice and snow lost from a glacier by the processes of melting, sublimation, wind erosion, and iceberg calving.

A horizon   The uppermost layer of a soil, containing organic material and leached minerals.

alpine glacier   See *valley glacier.*

amphibolite   A mostly nonfoliated metamorphic rock consisting primarily of amphibole and plagioclase feldspar.

andesite   A volcanic rock type intermediate in composition between rhyolite and basalt. The intrusive equivalent of diorite.

aphanitic texture   A fine-grain rock texture.

acquiclude   A stratum with low permeability that acts as a barrier to the flow of groundwater. Also called a *confining layer.*

aquifer   A permeable formation that stores and transmits groundwater in sufficient quantity to supply wells.

arete   The sharp, jagged crest along the divide between glacial cirques, resulting from the headward erosion of the walls of adjoining cirques.

artesian well   A well that is drilled into confined groundwater, causing the water under pressure to rise above the aquifer.

asthenosphere   The weak layer below the earth's lithosphere.

barrier island   A long, narrow island parallel to the shore, composed of sand and built by wave action.

basalt   A dark gray to black dense- to fine-grained igneous rock that consists of basic plagioclase, augite, and usually magnetite.

basaltic magma   Molten volcanic material composed of basalt.

base level   The level below which a stream cannot erode; usually sea level, but sometimes locally the level of a lake or resistant formation. For a river, the level of the body of water into which the river flows. Also, a sort of energy equilibrium at which the potential energy reaches a local or regional minimum.

bed load   The sediment that a stream moves along the bottom of its channel by sliding, rolling, and bouncing (saltation).

bedrock   Solid rock just below the surface.

B horizon   The intermediate layer in a soil, below the A horizon and above the C horizon, consisting of clays and oxide materials.

biochemical sediments   New chemical substances and deposits formed by organic processes.

boudinaged   Thinned.

breccia, volcanic   A pyroclastic rock in which all fragments are more than 2 mm in diameter.

brittle material   Material that, under pressure, deforms little or not at all until it suddenly breaks.

caldera   A steep-sided depression caused by the collapse of a volcano.

carbonate rock   A sediment or sedimentary rock formed from the accumulation of carbonate minerals precipitated organically or inorganically. Rocks are chiefly limestone and dolostone.

cataclasite   A rock changed by cataclastic metamorphism.

cataclastic metamorphism   High-pressure metamorphism occurring primarily by the crushing, shearing, and recrystallization of rock during tectonic movement and resulting in the formation of very fine grained rock.

channel   The trough through which lava flows in a stream down the slope of a volcano.

chemical sedimentary rock   A sediment or sedimentary rock that is formed at or near its place of deposition by chemical precipitation, usually from seawater.

chemical weather   The set of all chemical reactions that can act on rock exposed to water and the atmosphere and so dissolve the minerals or change them to more stable forms.

C horizon   The lowest layer of a soil, consisting of fragments of rock and their chemically weathered products.

cinder cone   A steep, conical hill built up about a volcanic vent and composed of pyroclastic, or igneous, rock fragments expelled from the vent by escaping gases.

cirque   The head of a glacial valley, usually with the form of one half of an inverted cone. The upper edges have the steepest slopes, approaching the vertical, and the base may be flat or hollowed out. The base is commonly occupied by a small lake or pond after deglaciation.

clastic particles   Mineral fragments broken off from the parent rock.. Also, particles of geological matter such as mud, sand, and gravel created by the mechanical breakdown of rocks.

clay   Any of a number of hydrous aluminosilicate minerals with sheetlike crystal structure, formed by the weathering and hydration of other silicates. Also, any mineral fragments smaller than 0.0039 mm.

cleavage   The inclination of a crystal or rock to split along definite planes.

climate   The average precipitation and temperature over a significant amount of time, usually about 40 years.

composite volcano   A volcano that expels both lava and pyroclasts. Also known as a *stratovolcano*.

compressive stress state   The force that exists between two of the earth's plates that are converging.

confined aquifer   An aquifer overlain by relatively impermeable strata (aquicludes), thereby causing the water to be contained under pressure.

contact metamorphism   Changes in rocks that result from contact with a hot igneous substance such as lava flow.

continental crust   The earth's crust that supports the mass of the continents.

continental glacier   A continuous, thick glacier covering more than 50,000 km^2 and moving independently of minor topographic feature. Also known as an *icecap*.

convergent plate boundaries   The edges of two of the earth's plates that have come together.

country rock   The rock into which an igneous rock intrudes.

crater   A circular depression caused by the activity of a volcano.

crust   The outermost layer of the lithosphere, consisting of relatively light, low-melting-temperature material.

delta, river   A sedimentary deposit formed where a river meets a large, open body of water.

density   The property of a mineral directly related to the atomic weight of its ions and to how tightly packed these atoms are bonded.

dike   An intrusive igneous rock that cuts through strata of surrounding rock.

diorite   A plutonic rock with a composition intermediate between granite and gabbro; the extrusive equivalent of andesite.

discharge   The volume of flow per unit of time, measured in cubic meters per second (m^3/s) or cubic feet per second (ft^3/s).

dissolved load   In a river or stream, the ions that have been broken down from their original forms as part of a solid and that are now chemically part of the water. The dissolved load is part of chemical weathering.

divergent plate boundaries   The edges of two of the earth's plates that have moved away from each other.

dolerite   A coarse-grained basaltic rock.

ductile material   A material that, under pressure, deforms gradually.

dune   An elongated mound of sand formed by wind or water.

dynamic metamorphism   Changes in rock that result from high pressure and low temperature such as those found at the surface of a fault.

earthquakes   The violent motion of the ground caused by the passage of seismic waves radiating from a fault along which sudden movement has taken place.

effluent stream   A stream or portion of a stream that receives some water from groundwater discharge because the stream's elevation is below the groundwater table.

elasticity   The capability of materials to regain their original shape after deformation.

epicenter   The point on the earth's surface directly above the focus of an earthquake.

erosion   The set of all processes by which soil and rock are loosened and moved downhill or downwind.

esker   A glacial deposit of sand and gravel in the form of a continuous, winding ridge, formed from deposits of a stream flowing beneath the ice.

extrusive igneous rock   An igneous rock formed from lava or from other products of volcanic material spewed out onto the surface of the earth.

evaporite sediments   Rocks that are precipitated inorganically from evaporating seawater and from water in arid-region lakes that have no river outlet.

fault   A fracture or break in the earth's crust.

felsic rock   An igneous rock that is light colored. Rock that is poor in iron and magnesium and contains abundant feldspars and quartz.

firn   Old, dense, compacted snow.

fissility   The capability of being split or divided along natural planes of cleavage.

fissure eruption   A volcanic eruption emanating from an elongate fissure rather than a central vent.

flash flood   A local flood of great volume and short duration generally resulting from heavy rainfall.

floodplain   A level plain of stratified, unconsolidated sediment on either side of a stream, submerged during floods and built up by silt and sand carried out of the main channel.

focus, earthquake The point along a fault within the earth at which a rupture occurs. The point of origin of an earthquake. See also *epicenter*.

fold A bent or warped stratum, or sequence of strata, that was originally horizontal, or nearly so, and was subsequently deformed.

footwall In an inclined fault plane, the surface or block of rock that lies below the fault plane.

foliation A set of flat or wavy planes in metamorphic rock produced during structural deformation.

fracture The irregular breaking of a crystal along a surface not parallel to a crystal face.

gabbro A black, coarse-grained, intrusive igneous rock, composed of calcic feldspars and pyroxene; the intrusive equivalent of basalt.

geodetic variations Changes in the level of the land relative to some base level.

geothermal energy Energy generated by using the heat energy of the earth's crust.

gneiss A coarse-grained regional metamorphic rock that shows banding and parallel alignment of minerals.

granulite A nonfoliated, high-grade regional metamorphic rock with coarse interlocking grains, generally formed under high pressure and temperature. Usually granular in texture.

greenhouse effect A global warming effect in which carbon dioxide and water vapor absorb infrared radiation from the earth's surface and radiate it back to the surface.

groundwater The mass of water in the ground (below the saturated zone) occupying the total pore space in the rock and moving slowly where permeability allows.

half-life The amount of time it takes for half of a radioactive material to become stable.

hanging wall In an inclined fault plane, the surface or block of rock that lies above the fault plane.

hardness, mineral The relative resistance of a mineral to scratching.

headwaters The source of a stream or river.

high-grade rocks Metamorphic rocks formed under the higher temperatures and pressures of deeper crustal regions.

hornblende A mineral that is the common dark variety of aluminous amphibole.

hornfels A high-temperature, metamorphic rock of uniform grain size showing no foliation. Usually formed by contact metamorphism.

hydraulic gradient The slope of an aquifer.

hydrologic cycle   The cyclical movement of water from the ocean to the atmosphere, through rain to the surface, through runoff and groundwater to streams, and back to the sea.

ice cap   See *continental glacier*.

igneous rock   A rock formed by the solidification of magma.

index minerals   The characteristic minerals that define metamorphic zones formed under a restricted range of pressures and temperatures.

influent stream   A stream or portion of a stream that recharges groundwater through the stream bottom because its elevation is above the groundwater table.

intrusive igneous rock   Igneous rock that forced its way into a country rock while it was in a molten state. Also known as *intrusion*.

island arc   A linear- or arc-shaped chain of volcanic islands formed at a convergent plate boundary area.

joint, fault   A large and relatively planar fracture in a rock across which there is no relative displacement of the two sides.

karst topography   An irregular topography characterized by sinkholes, caverns, and lack of surface streams; formed in humid regions because an underlying carbonate formation has been riddled with underground drainage channels that capture surface streams.   (Named for a region in northern parts of the former Yugoslavia with an irregular terrain of hills and many sinkholes.)

lahar   A mudflow of unconsolidated volcanic ash, dust, breccia, and boulders that occurs when pyroclastic or lava deposits mix with rain or the water of a lake, river, or melting glacier.

lattice   The arrangement of atoms in a crystal.

lava   Molten rock that has issued from a volcano or from a fissure in the earth's surface.

Liquefaction   The process of sediment becoming a fluid mass.

lithification   The chemical and physical diagenetic processes that find a harden a sediment into a sedimentary rock.

lithosphere   The outer, rigid shell of the earth situated above the asthenosphere and containing the crust, the uppermost part of the mantle, the continents, and the plates.

loess   An unstratified, wind-deposited, dusty sediment rich in clay minerals.

low-grade rocks   Metamorphic rocks formed under the lower temperatures and pressures of shallower crustal regions.

luster   The general quality of the shine of a mineral surface.

mafic mineral   A dark-colored mineral rich in iron and magnesium and relatively poor in silica (for example, pyroxene, amphibole, or olivine).

mafic rock   A rocks rich in mafic minerals.

magma   Volcanic molten rock material within the earth.

mass wasting   The downslope movement of soil, sediment, or rock.

meander   A broad, semicircular curve in a stream that develops as the stream erodes the outer bank of a bend and deposits sediment (as point bars) against the inner bank.

metamorphic rock   A rock produced deep in the earth where it is subjected to high pressure and temperature that causes it to change its chemical composition, shape, crystallization, texture, or mineralogy while remaining in a solid state.

metamorphism   Change in the minerals and textures of rock caused by changes in chemistry, temperature, or pressure.

midocean ridge   A mountain range extending from the ocean floor. Also, a segment of the Mid-Atlantic Ridge, which is also called the Oceanic Ridge and the Mid-Ocean Ridge.

mineral   A naturally occurring, inorganic, crystalline solid with a specific chemical composition.

moraine   A glacial deposit of till left at the margins of an ice sheet. Subdivided into ground moraine, lateral moraine, medial moraine, and end moraine.

mylonite   A very fine grained metamorphic rock commonly found in major thrust faults and produced by shearing and rolling during fault movement.

normal fault   A steeply inclined fault in which the hanging wall has moved downward in relation to the footwall.

nué ardentes   Flows of molten ash and pyroclastic fragments.

oblique slip or transform plate boundary   An inclined fault plane that exists between two plates that are moving parallel to each other.

oceanic crust   The earth's crust that supports the mass of the ocean basins.

oceanic ridge   A mounded ridge that runs continuously through all the major ocean basins and that is formed by the accumulation of volcanic materials that erupt through the area between the boundaries of divergent plates.

O horizon   Organic layer of soil at the surface.

olivine   A mineral that is a complex silicate of magnesium and iron, usually greenish in color.

outwash   A sediment deposited by meltwater streams emanating from a glacier.

ozone A molecule (O₃) that absorbs ultraviolet radiation in the stratosphere but creates smog when it forms near the earth's surface.

peat Rich, organic vegetation composed of more than 50 percent carbon.

pedalfer soil A common soil type in temperate regions, characterized by an abundance of iron oxides and clay minerals deposited in the B horizon by leaching.

pedocal soil A common soil type in arid regions, characterized by accumulation of calcium carbonate in the B horizon.

perched water table The upper surface of an isolated body of groundwater that is perched above and separated from the main body of groundwater by an aquiclude.

permeability The ability of a formation to transmit groundwater or other fluids through pores and cracks.

phaneritic texture A large-crystal, coarse-grain rock texture.

phenacrysts A large crystal surrounded by a finer matrix in an igneous rock.

physical weathering The set of all physical processes by which an outcrop is broken up into smaller particles.

plates Any of the large movable segments into which the earth's crust is divided.

plate tectonics The study of the activity of the 15 rigid plates into which the earth's crust is divided.

plutonic rock An intrusive igneous rock.

porosity The percentage of the total volume of a rock that is pore space (not occupied by mineral grains).

precipitate To drop out of a saturated solution as crystals. The crystals that drop out of a saturated solution.

P waves The primary or fastest wave traveling away from a seismic event through the solid rock and consisting of a train of compressions and dilations of the material.

pyroclasts Igneous rocks expelled by volcanic activity.

pyroxene Any of a group of igneous-rock-forming silicate minerals that contain calcium, sodium, magnesium, iron, or aluminum. These minerals usually occur in short prismatic crystals or massive form and vary in color from white to dark green or black.

recharge In hydrology, the replenishment of groundwater, usually by infiltration of meteoric water through the soil.

regional metamorphism Changes in rocks and minerals imposed by high temperature and high pressure such as that exerted below the earth's surface.

regolith   The layer of loose, heterogeneous material lying on top of bedrock. Regolith includes soil, unweathered fragments of parent rock, and rock fragments weathered from bedrock.

rejuvenation, mountain   Renewed uplift in a mountain chain on the site of earlier uplifts, returning the area to a more youthful stage of the erosion cycle.

reverse fault   A fault inclined at an angle greater than 45 degrees at which the hanging wall has moved up and over the footwall.

rhyolitic lava   A fine-grained volcanic or extrusive equivalent of granite, light-brown to gray and compact.

rhyolitic rock   An igneous rock composed of granite.

rhyolitic volcanism   Volcanic activity in which rhyolitic lava is expelled.

rift valley   A fault trough formed at a divergent plate boundary area or other area of tension.

rift zone   A normal fault area in which new crust is being created by divergent plate activity.

Ring of Fire   An arc of volcanoes in the Pacific Ocean, including Japan, Indonesia, and other island arcs, that is being formed by the subduction of oceanic crust under oceanic crust.

river   A general term for a relatively large stream or the main branches of a stream system.

rock flour   A glacial sediment of extremely fine (silt- and clay-sized) ground rock formed by abrasion of rocks at the base of a glacier.

schist   A metamorphic rock characterized by strong foliation or schistosity.

schistosity   Micas and amphiboles resulting from metamorphism.

seafloor spreading   The activity of two divergent plates that are pulling away from each other.

sediment   Any of a number of materials deposited at the earth's surface by physical agents such as wind, water, and ice; by chemical agents such as by precipitation from oceans, lakes, and river; or by biological agents such as by living or dead organisms.

sediment load   The distribution of materials carried by a river or stream as it flows from the headwaters into a base-level body of water.

sedimentary rock   A rock formed by the accumulation and cementation of mineral grains by wind, water, or ice transportation to the site of deposition or by chemical precipitation at a particular site.

seismicity   The worldwide or local distribution of earthquakes in space and time. Also a general term for the number of earthquakes in a unit of time.

seismic surface wave   A seismic wave that follows the earth's surface only with a speed less than that of S waves.

seismic wave   An elastic wave produced by earthquakes or explosions. Also the vibration caused by an earthquake.

shear stress state   The force that exists between two of the earth's plates that moving parallel to each other.

shield volcano   A type of volcano shaped like a shield with a broad circumference that has been formed by thousands of thin, frequently occurring basaltic lava flows. Also, a large, broad volcanic cone with very gentle slopes built up by nonviscous basaltic lava flows.

shoreline   The boundary where the land comes in contact with a body of water such as an ocean or lake.

silica   Silicon dioxide, SiO_2. A compound that occurs in crystalline, amorphous, and impure forms as in, respectively, quartz, opal, and sand.

silicate rock   An igneous or metamorphic rock.

sill   A horizontal, tabular igneous intrusion running between parallel layers of bedded country rock.

silt   Loose sedimentary material with rock particles usually 0.062 mm or less in diameter. Also, soil containing 80 percent or more of such silt and less than 12 percent of clay.

siltstone   A clastic rock that contains mostly silt-sized material, from 0.0039 to 0.062 mm.

sinkhole   A steep depression caused in karst topography by the dissolution and collapse of subterranean caverns in carbonate formations.

slip   The distance that two of the earth's plates slide past each other along a fault line during an earthquake. Also, the amount of parallel motion of tectonic plates that occurs during an earthquake.

slip, fault   The motion of one face of a fault relative to another.

soil   The surface accumulation of sand, clay, and humus that composes the regolith, but excluding the larger fragments of unweathered rock.

solubility, mineral   The extent to which a mineral can dissolve in water. The amount of the mineral dissolved in water when the solution reaches the saturation point.

solum   The altered layer of soil above the parent material that includes the A and B horizons.

strain Accumulated energy stored in rock formations locked into position on either or both sides of a fault plane.

strategic materials Nonrenewable resources that are important to industry.

stratovolcano See *composite volcano*.

streak The color a mineral leaves behind when it is scraped against a streak plate.

stress Pressure or force.

stress drop The release of energy accumulated and stored in the rocks locked into place on either or both sides of a fault plane.

stress state The existence of force between two objects.

strike The direction of the line of force that is creating a fault.

strike-slip fault A high-angle fault in which the plate movement is parallel to the fault line (*strike*) and there is little or no vertical displacement.

subduction boundaries The edges of two plates that have come together with the edge of one plate slipping under the edge of the other.

S wave The secondary seismic wave, which travels more slowly than the P wave and consists of elastic vibrations transverse to the direction of travel.

tectonics The study of forces within the earth that have led to the shaping of the earth's crust.

tensional stress state The force that exists between two of the earth's plates that are diverging.

thrust fault A fault inclined less than 45 degrees at which the hanging wall has moved up and over the footwall.

till An unstratified and poorly sorted sediment containing all sizes of fragments from clay to boulders, deposited by glacial action.

tombolo A sand or gravel bar connecting an island with the mainland or another island.

topography The shape of the earth's surface, above and below sea level. The set of landforms in a region. The distribution of elevations.

transcurrent fault A strike-slip fault.

transect A sample area of soil or vegetation, usually cut in a long continuous strip.

transform boundaries The edges of two plates that are gliding by each other laterally.

trenches The lowest areas on the earth's surface, all of which are in the deepest parts of the world's oceans.

valley glacier A glacier that is smaller than a continental glacier, or an icecap, that flows mainly along well-defined valleys in mountainous regions. Also known as an *alpine glacier*.

volcano A hill or mountain that forms from the accumulation of matter that erupts at the surface.

upwelling Rising to the surface and flowing outward.

viscosity A measure of a liquid's resistance to flow.

weathering The set of processes that decay and break up rock, by a combination of physical fracturing and chemical decomposition.

Answers to End-of-Unit Review Questions

Module One: Materials and Processes

1. Mineralogy

1. Given an example of a mineral in the oxide group. The CD shows hematite, magnetite, and corundum.

2. What is luster? The appearance of a mineral surface when reflecting light.

3. "One atom of carbon and three atoms of oxygen ions" describes what type of mineral? Carbonates.

4. In the earth, what class(es) of minerals tend to make up igneous rocks? Silicates.

2. Igneous Rocks

1. When magma erupts onto or near the earth's surface and cools rapidly, what type of rock texture will the cooled rocks have? What type of texture do they have if the magma intrudes into deep rock inside the earth? Fine grained; course-grained.

2. What island chain is offered as an example of a site of rapidly-cooling mafic magma? The Hawaiian Islands.

3. In rift zones at divergent plate boundaries, what often happens to the magma that does not reach the surface and erupt in a basaltic flow? (see "mafic," "slow") It often forms sills or dikes.

3. Metamorphic Rocks

1. What three factors cause regional metamorphism? Heat, pressure, and directed stress.

2. What causes rocks to change in a dynamic metamorphic situation? Fault activity.

3. How do intrusive igneous rocks cause contact metamorphism to occur? The heat in the crystallizing magma in the intrusion causes recrystallization of the country rock in a thin zone around the intrusive body.

4. Physical Properties

1. Which stress state tends to promote sliding along a plane? Shear.

2. When brittle rock fractures, what happens to the stored elastic strain energy? It is released as microseismic or seismic waves.

3. If a strong material behaving in a brittle fashion were subjected to shear stresses, what type of fracture would result? A vertical fracture.

4. If a strong material behaving in a ductile fashion were subjected to shear stress, what would the effect be on the surrounding shear zone? The material would display recrystallization of mineral phases and evidence of flow banding.

5. Rock Cycle

1. What is lithification? A process in which the grains of buried rock material are cemented together by the percolation of groundwater through it, forming sedimentary rock.

2. How are rocks changed in uplifted areas exposed to weathering? Wind, water, and gravity weather away the rocks to form sediment.

6. Sedimentary Rocks

1. How is rock salt formed? By the evaporation of water that is carrying sodium chloride in solution.

2. How do organisms form sedimentary materials? They use calcium carbonate or silica available in their environment to build their shells or skeletons. When they die, their shells are deposited on the seafloor where they are compacted and lithified.

3. What characteristic distinguishes glacial sedimentary deposits from other types of transported materials? The heterogeneity of the types of rocks.

7. Soils

1. Where is the O horizon found in a soil profile? At the surface.

2. What would the organic matter and mineral content be in the soil of an arid region? Low in organic matter, high in minerals.

3. What property of clay makes its presence an impediment in the soil-forming process? Its poor permeability.

8. Weathering

1. What effect do frequent freezing and thawing periods have on rocks? Increased rate of rock fragmentation.

2. What types of weathering result from wind and gravity in areas where surface water is present? Accelerated physical and chemical weathering.

3. What two factors inhibit chemical weathering in polar climates? The cold and lack of liquid water.

4. What factor inhibits chemical weathering in arid climates? The lack of water.

5. What two climatic factors affect the rate of weathering of rocks? The temperature and humidity.

Module Two: Earthquakes and Volcanoes

1. Plate Tectonics

1. The Appalachian Mountains are an example of a plate collision at what type of plate boundary? Convergent.

2. What is the area called when plates pull apart from each other at divergent boundaries? A rift zone.

3. Name a geologic process that might occur at a subduction zone. Andesitic volcanism; low-temperature, high-pressure metamorphism; high-grade regional metamorphism

4. What type of tectonic plate activity has caused the San Andreas Fault system in southwestern California? Two plates are sliding laterally past one another.

2. Stress States

1. In what type of tectonic activity do tensional stresses occur? In the upper part of divergent plate boundaries, or rift zones.

2. In a brittle regime, high-stress-drop situation, what three forms of elastic strain energy are released? Kinetic, thermal, and elastic energy.

3. In a brittle regime, what types of faults result from compressive stresses? Reverse faults or low-angle thrust faults.

4. In a normal fault, in what direction does the hanging-wall block move in relation to the footwall block? Down.

5. Why does the ductility of rock increase with depth in the earth's crust? Both temperature and pressure increase with depth. The higher the temperature and pressure, the greater the ductility of the rock.

3. Magma Types

1. What effect does the rate of cooling have on the crystallization of melted rock material? Slow cooling results in large crystals; quick cooling results in small crystals.

2. What are shield volcanoes? Low-slope, large area volcanoes.

3. What is the viscosity of the three basic types of magma? Basaltic, low viscosity; andesitic, intermediate viscosity; and rhyolitic, high viscosity.

4. What type of volcano results from eruptions of basaltic magma? Shield volcanoes.

5. What type of volcano results from eruptions of andesitic magma? Composite volcanoes (also called stratovolcanoes).

6. What type of volcano results from eruptions of rhyolitic magma? Cinder cones.

4. Explosivity

1. How can volcanism sometimes cause air pollution? Eruptions release particulate matter and gases into the atmosphere.

2. How can volcanism actually benefit human activity? Volcanism can enrich soils for agriculture and be a source of geothermal energy.

3. How can volcanism sometimes cause changes in climate? Particle matter released from eruptions can be so great that it absorbs ultraviolet radiation from the sun and filters the reflection of energy from the surface of the earth.

4. How can volcanism be hazardous to human safety? Lava, mud, pyroclastic, and molten ash flows can engulf structures and people, and toxic fumes or very hot fiery gas clouds can poison people quickly.

Module Three: Surface Processes

1. Rivers

1. What type of sediment load is composed of coarser-grained sediment that is too heavy to be suspended? Bed load.

2. What effect does mountain building have on river discharge? It raises the level of the landscape over which the river flows relative to the base level of the river. This increases the potential energy of the river.

3. What effect does a lowering of the sea level have on river discharge? The volume of water in the river decreases. The rate of erosion and weathering may increase as a result.

2. Coastal Processes

1. What happens at the shoreline when the sea level lowers? The beach front moves forward, and the shelf and platform migrate with it.

2. What happens at the shoreline when the sea level rises? The water advances onto land that was once above water.

3. During mountain building, what causes a coastline to prograde, or move outward into the sea? Increased weathering and erosion produce a great deal of sediment, which is deposited on the coastline.

4. Give an unintended effect of a jetty. The collection of sand on one side of the jetty and the lose of sand on the other side.

3. Wind

1. What causes air to circulate from regions of high pressure to regions of low pressure? The differential heating of the earth's surface.

2. In what direction does polar cold air flow? Toward the low-pressure area in the lower latitudes.

3. In what way do blowout dunes differ in appearance from barchan dunes? The points of blowout dunes face the wind direction.

4. Glaciers

1. What force causes glaciers to move? Gravity.

2. What is till? Unsorted, nonstratified, unconsolidated sediment that has been transported and deposited by a glacier.

3. What happens when glacial ice reaches a thickness of 30-40m? It can no longer support its own weight and beings to flow.

Module Four: Groundwater

1. Saltwater Intrusion

1. How many states in the United States have coastal frontage? 28.

2. What is the primary source of freshwater aquifer recharge in coastal zone developments in stages 1 through 4? Cesspools.

3. What are the two primary strategies that are successful in lessening saltwater intrusion? (a) Regulating the withdrawal of freshwater, (b) using recharge basins to collect storm runoff from residential and industrial areas and highways.

2. Darcy's Law

1. The material type for an aquifer is sand. The cross-section area is 1 and the hydraulic gradient is 1/3. What is the rate of groundwater flow through the aquifer? 495.

2. Through which material would groundwater flow faster, clay or gravel? Gravel.

3. Porosity and Permeability

1. Gravels have excellent porosity and permeability. Why, then, do they make poor aquifer materials? Water flows through this material so easily that it has no storage capability.

2. How does sand hold water? With surface tension.

3. The porosity and permeability of sandstone depend on what factors? Cementation and jointing.

Module Five: Natural Hazards

1. Volcanoes

1. Of the three volcanic types represented on the CD, which one poses the greatest threat of producing lahars? Cinder cone volcanoes.

2. When did Mount St. Helens last erupt? 1986.

3. Which volcanic danger did the appearance of blue flames at Mount St. Helens in march 1980 indicate to scientists? Gas emissions.

2. Flooding

1. What is the definition of flooding? The rapid runoff of the overland flow of water.

2. What four factors strongly affect the rate of surface runoff? The steepness of the slope, vegetation density, permeability of the surfacial material, and rainfall intensity and duration.

3. List some of the impermeable surfaces found in urbanized areas. Homes, buildings, parking lots, streets, sidewalks.

3. Earthquakes

1. What type of earthquake damage effect probably poses the most danger to man-made buildings? Swaying.

2. Earthquake activity at transform boundaries is relatively rare because this type of tectonic activity is rare, but there are two such seismically active areas in the world. Where are these two areas? The San Andreas and North Anatolian fault systems.

3. What is the name of the term for changes in the elevation of ground near earthquakes due to dialation of deformed rock? Geodetic variation.

4. Mass Wasting

1. What type of "materials" situation is the most stable? Upslope.

2. How does vegetation offset the risk of mass wasting? Vegetation tends to anchor soil and sediment and also to remove water from the material.

3. When soil materials predominate, a slow form of mass wasting is a potential danger. What is the term for this slow form of mass wasting? Creep.

Module Six: Waste Disposal

1. Waste Types

1. What is the second leading source of commercial solid waste in the U.S.? Mining.

2. What sorts of hazardous wastes are generated by farming activities? Pesticides and herbicides.

3. Why is radioactive waste more difficult to dispose of than other types of waste? It has a long half-life, that is, the amount of time it takes for half of a radioactive material to become stable.

2. Site Selection

1. What hazard would an earthquake pose to a radioactive waste storage facility? An earthquake may cause a breaching of the containment vaults.

2. What hazard would flooding pose to a radioactive waste storage facility? Should a breaching of the containment vaults occur, flooding would spread radioactive waste away from the storage facility quickly and uncontrollably.

3. What is a favorable factor for radioactive waste storage relating to the groundwater at Hanford? The water table is low.

4. What possible dangers would there be in moving the radioactive storage bunkers at Hanford above ground? It would make them more susceptible to terrorist activity.

Module Seven: Resources and Sustainability

1. Strategic Materials

1. According to the U.S. Geological Survey, how long will the world's petroleum reserves last? Only until the end of the twenty-first century.

2. According to the U.S. Geological Survey, how long will the world's natural gas reserves last? Until into the twenty-second century.

3. What industries use the most chromium? The automobile and appliance industries.

4. Name a strategic material that South Africa produces in major quantities? Platinum and chromium (as well as cobalt, magnesium, gold, and diamonds.

5. What strategic material is found in the Lake Superior region of the U.S. and Canada? Iron.

2. Resources Politics

1. What does the term net producer *mean?* A country that uses less of the strategic materials than it produces.

2. What would the effect be on the rest of the world if the United States were removed as a supplier and consumer? The loss of the United States would greatly reduce the consumption of fuels and materials, but it would also create economic decline among the major producing countries.

3. What percentage of the world's fossil fuels does the U.S. produce? Less than five percent.

4. Does Germany consume more energy than Great Britain? No, Great Britain consumes eleven percent versus five percent by Germany.

3. Public Policy

1. What percentage of current U.S. energy needs is met with petroleum and natural gas? Seventy percent.

2. How did the public respond to the rationing of fuel during World War II? The public accepted rationing fairly well during the war; however, some people tried to circumvent the rationing by illegal means such as purchasing fuel through a black market.

3. What percentage of the U.S. population uses public transportation and carpooling? Less than five percent.

4. What is currently the most expensive form of energy in the U.S. in terms of cost per kilowatt hour? Solar.

Module Eight: Field Trips

1. Landfills

1. What purpose does clay serve in a landfill? Clay-rich material is placed over the plastic liner and serves as an impermeable boundary.

2. What does the leachate monitoring well do with excess leachate? It pumps it from the well into a holding pool.

3. What is the final stage of landfill development? A conditioned clay cap covers the site and it is revegetated.

2. Faults

1. What is the technical term for the cracks in the parking lot at the former town hall? En echelon fractures.

2. Why is there apt to be lush vegetation in an active fault zone? Vegetation tends to thrive in fault zones due to the presence of increased groundwater in the highly fractured area.

3. Why do modular structures withstand earthquakes better than other types of structures? The modules can respond to the earthquake stress and move independently of each other.

3. Tectonics

1. What type of tectonic activity would the presence of hydrothermal deposits suggest? Igneous activity.

2. What makes the melange rock green? Serpentine.

3. How are pillow basalts formed? These formations occur when basaltic lava is cooled rapidly in ocean water and separates into pillow-shaped globules.

Correlation of CD-ROM to Freeman Textbooks

By Textbook Chapters

Environmental Geology Textbook

Environmental Geology: An Earth System Science Approach by Dorothy Merritts, Andrew De Wet, Kirsten Menking

Chapter 1: Introduction to Environmental Geosciences
➢ "Resources and Sustainable Development," pp. 17-19" — *Resources and Sustainability Module/all Units*

Chapter 2: Dynamic Earth Systems
➢ Chapter 2 Introduction, pp. 30-31 — *Natural Hazards Module/Unit 1: Volcanoes*
➢ "Where Soils Form," pp. 42-44 — *Materials and Processes Module/Unit 7: Soils*
➢ "Earth's Energy System," pp. 47-50" — *Resources and Sustainability Module/all Units*
➢ "Igneous Rocks," p. 53 — *Earthquakes and Volcanoes Module: Unit 3: Magma Types*
➢ "The Rock Cycle," pp. 53-55 — *Materials and Processes Module/Unit 2: Igneous Rocks; Unit 3: Metamorphic Rocks; Unit 5: Rock Cycle; Unit 6: Sedimentary Rocks*

Chapter 3: Geologic Time and Earth History
➢ Chapter 3 Introduction, p. 64; "Predicting the Geologic Stability of Yucca Mountain," pp. 82-83 — *Waste Disposal Module/all Units*
➢ "Predicting the Eruption of Mount St. Helens," pp. 67-68 — *Natural Hazards Module/Unit 1: Volcanoes*

Chapter 4: Lithosphere: The Rock and Sediment System
➢ Chapter 4 Introduction, pp. 92-93; "Plate Tectonics and the Rock Cycle," pp. 102-110 — *Earthquakes and Volcanoes Module/Unit 1: Plate Tectonics — Field Trips Module/Unit 3: Tectonics*
➢ "Lithosphere Materials: Elements, Minerals, and Rocks," pp. 93-107 — *Materials and Processes Module/Unit 1: Mineralogy*
➢ "Major Rock Groups and the Rock Cycle," pp. 100-101; "Distribution of Rock Types," pp. 111-118 — *Materials and Processes Module/Unit 2: Igneous Rocks; Unit 3 Metamorphic Rocks; Unit 4: Rock Cycle; Unit 6: Sedimentary Rocks*
➢ "Plate Tectonics and the Rock Cycle," pp. 102-110
➢ "Igneous Rocks," pp. 111-113 — *Earthquakes and Volcanoes Module/Unit 3: Magma Types*

Chapter 5: Lithosphere: Resources, Hazards, and Change
➢ General — *Materials and Processes Module/Unit 4: Physical Properties*
➢ General — *Earthquakes and Volcanoes Module/Unit 2: Stress States*

➤ "Rock and Mineral Resources," pp. 124-138" — *Resources and Sustainability Module/all Units*

➤ "Butte, Montana — From Boom Town to Superfund Site," pp. 132-133 — *Waste Disposal Module/all Units*

➤ "Volcanic Eruptions," pp. 138-139; "Crustal Deformation Associated with Earthquakes in the Cascades Region," pp. 153-154 — *Earthquakes and Volcanoes Module/Unit 1: Plate Tectonics*

➤ "Volcanic Eruptions," pp. 138-146 — *Earthquakes and Volcanoes Module/Unit 3: Magma Types; Unit 4: Explosivity; Natural Hazards Module/Unit 1: Volcanoes*

➤ "Earthquakes," pp. 146-154 — *Natural Hazards Module/Unit 3: Earthquakes; Field Trips Module/Unit 2: Faults*

Chapter 6: Soil Systems and Weathering

➤ General — *Materials and Processes Module/Unit 7: Soils*

➤ "Soil Erosion Hazards and Soil Conservation," pp. 179-182 — *Surface Processes Module/Unit 3: Wind*

➤ "Mass Movement Hazards and Their Mitigation," pp. 182-188 — *Natural Hazards Module/Unit 4: Mass Wasting*

Chapter 7: The Surface Water System

➤ General — *Surface Processes Module/Unit 1: Rivers*

➤ Chapter 7 Introduction, pp. 194-195; "The Hazards of Flooding," pp. 206-216 — *Natural Hazards Module/Unit 2: Flooding*

➤ "Surface Resources and Protection," pp. 220-228 — *Waste Disposal Module/all Units; Field Trips Module/Unit 1: Landfills*

Chapter 8: The Groundwater System

➤ "Porosity and Groundwater Storage," pp. 235-237; "Permeability and Groundwater Storage," pp. 237-238 — *Groundwater Module/Unit 3: Porosity and Permeability*

➤ "Wells," pp. 239-242; "Darcy's Law and the Flow of Water and Contaminants in Rocks and Sediments," pp. 241-242 — *Groundwater Module/Unit 2: Darcy's Law*

➤ "Intrusion of Salt Water," pp. 249-250 — *Groundwater Module/Unit 1: Saltwater Intrusion*

➤ "Groundwater Pollution and Its Cleanup," pp. 250-256 — *Waste Disposal Module/all Units; Field Trips Module/Unit 1: Landfills*

Chapter 9: The Atmospheric System

There is no direct coverage of the material in this chapter on *Earth Matters*.

Chapter 10: The Ocean and Coastal System

➤ General — *Surface Processes Module/Unit 2: Coastal Processes*

➤ Chapter 10 Introduction, pp. 292-293; "Ocean Pollution," pp. 316-321 — *Waste Disposal Module/all Units*

Chapter 11: Energy and the Environment

➤ General — *Resources and Sustainability Module/all Units*

Chapter 12: Understanding Environmental Change
➤ "Indicators of Environmental Change," pp. 372-376 — *Surface Processes Module/Unit 4: Glaciers*
➤ "Arid Environments," pp. 373-374 — *Surface Processes Module/Unit 3: Wind*

Chapter 13: Tracing and Predicting Environmental Change
➤ "The Little Ice Age," pp. 399-400 — *Surface Processes Module/Unit 4: Glaciers*

Physical Geology Textbook

Understanding Earth, Second Edition, by Frank Press and Raymond Siever

Chapter 1: Building a Planet
➤ "Plate Tectonics: A Unifying Theory for Geological Sciences," pp. 14-20 — *Earthquakes and Volcanoes Module/Unit 1: Plate Tectonics*
➤ "Plate Tectonics: A Unifying Theory for Geological Sciences," pp. 14-20 — *Field Trips Module/Unit 3: Tectonics*

Chapter 2: Minerals: Building Blocks of Rocks
➤ General — *Materials and Processes Module/Unit 1: Mineralogy*

Chapter 3: Rocks: Records of Geologic Processes
➤ "The Rock Cycle," pp. 68-70; "Plate Tectonics and the Rock Cycle," pp. 70-71 — *Materials and Processes Module/Unit 5: Rock Cycle*

Chapter 4: Igneous Rocks: Solids from Melts
➤ General — *Materials and Processes Module/Unit 2: Igneous Rocks*
➤ "How Does Magma Form?" pp. 83-84; "Where Does Magma Form?" pp. 85-86 — *Earthquakes and Volcanoes Module/Unit 3: Magma Types*

Chapter 5: Volcanism
➤ General — *Earthquakes and Volcanoes Module/Unit 3: Magma Types*
➤ General — *Earthquakes and Volcanoes Module/Unit 4: Explosivity*
➤ "The Global Pattern of Volcanism," pp. 121-125 — *Earthquakes and Volcanoes Module/Unit 1: Plate Tectonics*
EM Unit: Volcanoes
➤ "Volcanism and Human Affairs," pp. 125-131 — *Natural Hazards Module/Unit 1: Volcanoes*

Chapter 6: Weathering and Erosion
➤ General — *Materials and Processes Module/Unit 8: Weathering*
➤ "Soil: The Residue of Weathering," pp. 152-158 — *Materials and Processes Module/Unit 7: Soils*

Chapter 7: Sediments and Sedimentary Rocks
➤ General — *Materials and Processes Module/Unit 6: Sedimentary Rocks*

Chapter 8: Metamorphic Rocks
➤ General—*Materials and Processes Module/Unit 3: Metamorphic Rocks*

Chapter 9: The Rock Record and the Geologic Time Scale
There is no direct coverage of the material in this chapter on *Earth Matters*.

Chapter 10: Folds, Faults, and Other Records of Rock Deformations
➤ "How Rocks Become Deformed,", pp. 247-248—*Materials and Processes Module/Unit 4: Physical Properties*
➤ "How Rocks Become Deformed," pp. 247-248—*Earthquakes and Volcanoes Moduel/Unit 2: Stress States*
➤ : "How Rocks Become Formed," 247-248; "How Rock Fracture," pp. 255-258; "Unraveling Geological History," 258-260—*Field Trips Module/Unit 2: Faults*

Chapter 11: Mass Wasting
➤ General—*Natural Hazards Module/Unit 4: Mass Wasting*

Chapter 12: The Hydrologic Cycle and Groundwater
➤ General—*Groundwater Module/all Units*
➤ "How Water Flows Through Soil and Rock," pp. 296-298—*Groundwater Module/Unit 3: Porosity and Permeability*
➤ "Balancing Recharge and Discharge," pp. 301-302—*Groundwater Module/Unit 1: Saltwater Intrusion*
➤ "The Speed of Groundwater Flows," pp. 303-304—*Groundwater Module/Unit 2: Darcy's Law*
➤ "Water Quality," pp. 308-311—*Waste Disposal Module/all Units; Field Trips Module/Unit 1: Landfills*

Chapter 13: Rivers: Transport to the Oceans
➤ General—*Surface Processes Module/Unit 1: Rivers*
➤ "Stream Valleys, Channels, and Floodplains," pp. 323-326; "The Development of Cities in Floodplains," p. 327—*Natural Hazards Module/Unit 2: Floods*

Chapter 14: Winds and Deserts
➤ General—*Surface Processes Module/Unit 3: Wind*

Chapter 15: Glaciers: The Work of Ice
➤ General—*Surface Processes Module/Unit 4: Glaciers*

Chapter 16: Landscape Evolution
There is no direct coverage of the material in this chapter on *Earth Matters*.

Chapter 17: The Oceans
➤ General—*Surface Processes Module/Unit 2: Coastal Processes*

Chapter 18: Earthquakes
➤ General—*Earthquakes and Volcanoes Module/Unit 1: Plate Tectonics*

➤ "Earthquake Destructiveness," pp. 472-481 — *Natural Hazards Module/Unit 3: Earthquakes*
➤ "The Oceans as a Deep Waste Repository," p. 450 — *Waste Disposal Module/all Units*
➤ "What Is an Earthquake?" pp. 460-462 — *Field Trips Module/Unit 2: Faults*
➤ "The Big Picture: Earthquakes and Plate Tectonics,: pp. 470-472 — *Field Trips Module/Unit 3: Tectonics*

Chapter 19: Exploring Earth's Interior
There is no direct coverage of the material in this chapter on *Earth Matters*.

Chapter 20: Plate Tectonics: The Unifying Theory
➤ General — *Earthquakes and Volcanoes Module/Unit 1: Plate Tectonics*
➤ General — *Field Trips Module/Unit 3: Tectonics*

Chapter 21: Deformation and the Continental Crust
➤ "Radioactive Waste Disposal," pp. 574-575; "Nuclear Energy Hazards," pp. 575-576 — *Waste Disposal Module/all Units*

Chapter 22: Energy Resources from the Earth
➤ General — *Resources and Sustainability Module/all Units*
➤ General — *Field Trips Module/Unit 1: Landfills*

Chapter 23: Mineral Resources from the Earth
➤ General — *Resources and Sustainability Module/all Units*

Chapter 24: Earth Systems and Cycles
There is no direct coverage of the material in this chapter on *Earth Matters*.